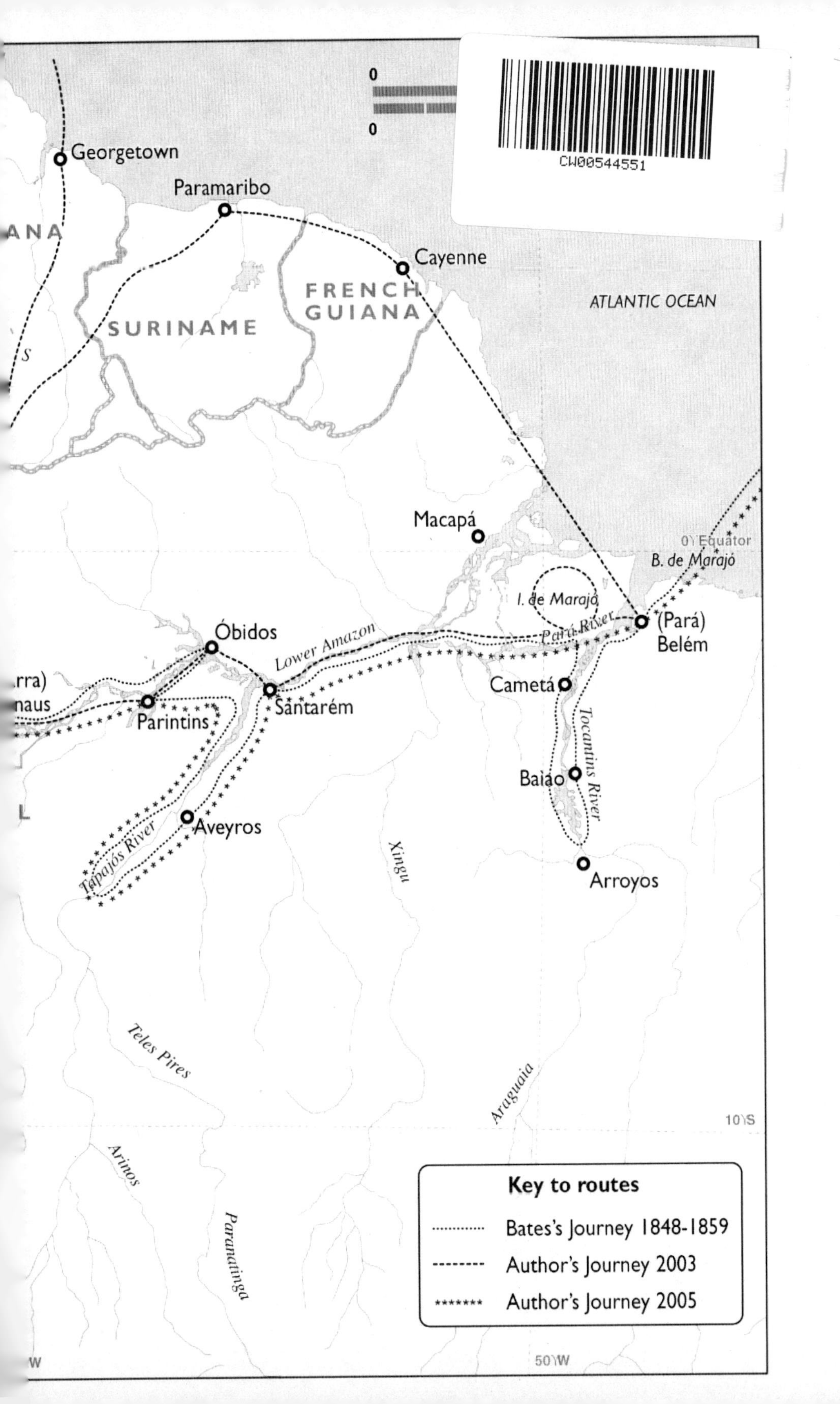

Georgetown
Paramaribo
ANA
Cayenne
FRENCH GUIANA
SURINAME
ATLANTIC OCEAN
Macapá
0° Equator
B. de Marajó
I. de Marajó
Óbidos
Lower Amazon
Pará River
(Pará) Belém
rra)
naus
Parintins
Santarém
Cametá
Tocantins River
Baiao
Aveyros
Tapajós River
Xingu
Arroyos
Teles Pires
Araguaia
10°S
Arinos
Paranatinga
Key to routes
Bates's Journey 1848-1859
Author's Journey 2003
Author's Journey 2005
50°W

The Butterfly Hunter

Henry Walter Bates FRS

1825 – 1892

To Roy and Margaret.

Anthony Crawforth

Anthony Crawforth

Published in 2009 by the University of Buckingham Press
Yeomanry House, Hunter Street
Buckingham MK18 1EG
United Kingdom

First Edition

ISBN 9780956071613

Printed in Great Britain by the MPG Books Group, Bodmin and King's Lynn

'We borrow the light of an observant and imaginative traveller, and see the foreign land bright with his aura; and we think it is the country that shines'

The Sea and the Jungle, H. M. Tomlinson. 1930

On his way back from Santa Maria de Belém do Grão Pará, in the Brazils.

Qualifications for a Traveller

If you have health, a great craving for adventure, at least a moderate fortune, and can set your heart on a definite object, which old travellers do not think impracticable, then travel by all means. If, in addition to these qualifications, you have scientific taste and knowledge, I believe that no career, in time of peace, can offer to you more advantages than that of a traveller. If you have not independent means, you may still turn travelling to excellent account; for experience shows it often leads to promotion, nay, some men support themselves by travel. They explore pastureland in Australia, they hunt for ivory in Africa, they collect specimens of natural history for sale, or they wander as artists.

Francis Galton, *The Art of Travel or Shifts and Contrivances available in Foreign Countries.* (London: John Murray, 1872) Preface to the fifth edition.

This is how it is

I like biography far better than fiction myself; fiction is too free. In biography you have your little handful of facts, little bits of a puzzle, and you sit and think, and fit 'em together this way and that, and get up and throw 'em down and say damn, and go out for a walk, and it's real soothing; and when done, gives an idea of finish to the writer that is very peaceful. Of course it's not really so finished as quite a rotten novel; it always has and always must have the incurable illogicalities of life about it.... Still, that's where the fun comes in.

Robert Louis Stevenson to Edmund Gosse, The Tusitala Edition, 35 volumes, Heinemann 1923-1924, *Letters:* vol V, 41.

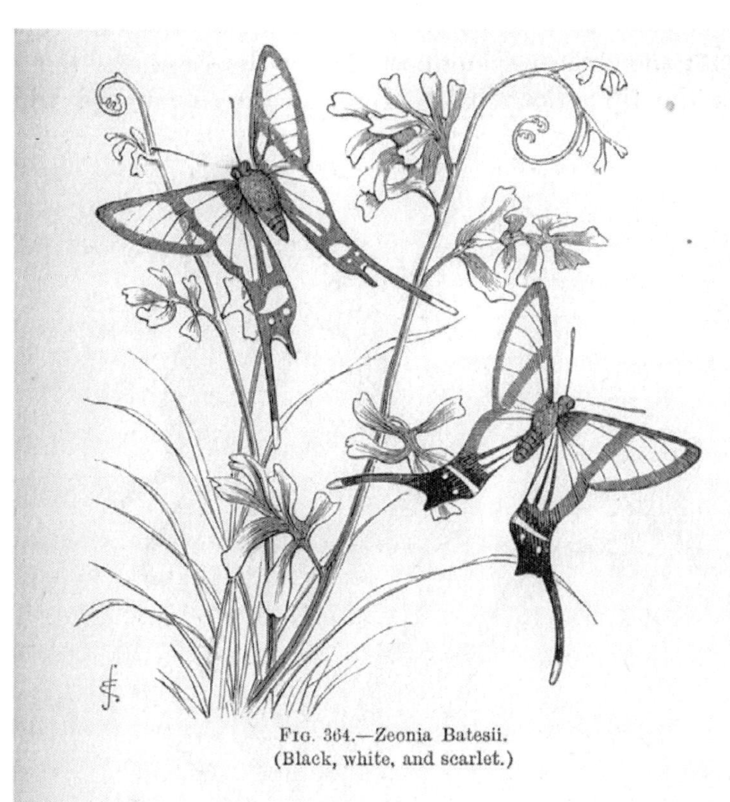

Fig. 364.—Zeonia Batesii.
(Black, white, and scarlet.)

Plate 1.
Zeonia Batesii

For Wendy

Plate 2.

Henry Walter Bates aged about 50, c 1875

Contents

H W Bates's Family Tree

Edward Bates	b c1640		
▼			
Robert Bates	b 10-5-1665	d 23-8 1728	
m 11-5-1699	**Avis Sclater**	b - ?	d - ?
▼			
Thomas Bates	b 15-10-1700	d 25-3-1759	
m 29-6-1729	**Elizabeth Dawson**	b - ?	d - ?
▼			
John Bates	b 25-4-1737	d - ?	
m 30-10-1764	**Mary Prior**	b - ?	d 20 -4 -1803
▼			
Robert Bates	b 19-1-1770	d 14-1-1837	
m 19-7-1792	**Mary Donnisthorpe**	b 1773	d 1-8-1819
▼			
Henry Bates	b 1794	d 24-8-1870	
m 6-5-1824	**Sarah Gill**	b - ?	d 19 -1-1860
▼			
Henry Walter Bates	b 8-2-1825	d 16-2-1892	
m 15-1-1863	**Sarah Ann Mason**	b 30 -7-1840	d 13 -10-1897
▼			

Alice b 1862 d 1891

Sarah b 1863 d 1929

Charles Henry b 1865 d 1901

Darwin b 1867 d 1938

Herbert Spencer b 1871 d 1958

Chronology

8th January **1823** - Alfred Russel Wallace born in Usk.
8th February **1825** - Henry Walter Bates born in Leicester.
Late **1836**/early **1837** - Wallace forced to withdraw from grammar school as formal education ends; moves to London to board with his older brother, John.
1838 - Bates's formal education ends but he later takes night classes at the Leicester Mechanics' Institute. Afterwards he is apprenticed to a hosiery manufacturer.
1841 - Wallace is associated with the Kington Mechanics' Institute.
1843 - Bates's first scientific publication on beetles for vol I of *The Zoologist.*1. p.115.
1844 - Bates meets Wallace who is hired as a master at the Collegiate School in Leicester.
1845 - Wallace's brother William dies. Wallace leaves Collegiate School and takes over his surveying business.
1845 – **1848** - Bates and Wallace become inseparable friends.
1848 - Bates's second scientific paper published, *Coleoptera of Burton on Trent, The Zoologist* VI. 1848 p 1997.
1848 - Bates and Wallace travel to Pará, at the mouth of the Amazon, to begin a joint natural history collecting expedition. First, they concentrate their collecting around (Pará) Belém and on the Rio Tocantins.
1849 - Bates and Wallace go their separate ways, Bates concentrating on the central and lower Amazon basin, Wallace on the Rio Negro and its tributaries.
1849 – May **1850** - Bates's travels to Manaus and then to Tefé, on the Upper Amazon.
Aug – Dec **1849** - Wallace travels independently to Manaus and then in **1851** travels up the Rio Negro to the Uaupés river.
1850 - Bates, Wallace, and his younger brother Herbert Wallace meet in Manaus
1851 - Bates returns to and nurses Wallace's dying brother in (Pará) Belém. Wallace is out of contact.
Late **1851** - Bates sets up his base in Santarém and remains there until **1855.**
1852 - Wallace leaves South America for return to England, on 6th August, his ship burns and sinks and ten days later he is rescued at sea, barely surviving the ordeal.
1852-**1854** - Wallace is primarily London-based.

1853 - Wallace publishes *Palm Trees of the Amazon* and *A Narrative of Travels on the Amazon and Rio Negro.*
1854 - Wallace leaves England for the Malay Archipelago to begin a new natural history collecting expedition.
1855 - Bates moves to (Ega) Tefé, establishes his base there until **1859** and explores as far west as São Paulo de Olivença.
1854-1862 - Wallace is collecting in the Malay Archipelago & communicating his ideas and findings with Bates.
1858 - Wallace writes *On the Tendency of Varieties to Depart Indefinitely from the Original Type* and sends it off to Charles Darwin for comment (Feb.)
1858 - Wallace's and Darwin's papers on natural selection are presented at a meeting of the Linnean Society. (July)
1859 - Bates leaves (Ega) Tefé and returns to (Pará) Belém (17.3.59) and England (2.6.59) after eleven years in Amazonia.
1860 - Bates's mother dies.
1862 - Bates marries Sarah Ann Mason.Wallace returns to England.
1862 - Bates publishes *Contributions to an Insect Fauna of the Amazon Valley. Lepidoptera: Heliconiidae* in the *Transactions of the Linnean Society*, developing theories on geographical distribution and the basis of mimetic protection.
1863 - Bates publishes his *The Naturalist on the River Amazons*, in two volumes.
1864 - The revised edition of *The Naturalist on the River Amazons* is published.
1864-1892 - Bates is appointed assistant secretary of the Royal Geographical Society.
1866 - Wallace marries Annie Mitten, daughter of botanist friend William Mitten.
1868-1869 & **1878** - Bates appointed president of the Entomological Society of London.
1869 - Wallace's *The Malay Archipelago* is published.
1869 - Wallace decides the human species cannot be fully explained by the theory of evolution.
1870 - Bates's father dies.
1871 - Bates elected a fellow of the Linnean Society of London.
1872 - Bates awarded the Order of the Rose by Dom Pedro II of Brazil.
1870 – 1872 - Wallace is President of the Entomological Society of London.
1876 - Wallace elected President, Section D (Biology) of the British Association for the Advancement of Science.
1881 - Bates is elected fellow of the Royal Society of London.
1882 - Death of Charles Darwin.
1892 - On 16th February Bates dies at his London home.

Illustration acknowledgements

Front cover details from plate VIII, *Beautiful Butterflies of the Tropics, How to collect them,* by Arthur Twidle, R.T.S, London 1926. [Bates approved this plate as both he and Wallace were involved in thr preparation of Twidle's Book]. Plates 1, 18, 19, 20, 21 emgraved by J.b. Zwecker 1892 for *Insects Abroad* by J.G. Wood. Plate 2, by J. Thompson 1892, Plates 10 and 12 by J.B. Zwecker from Bates's *The Naturalist on the River Amazons* 1863. Plate 3, by Jean-Baptiste Debret from the album *Voyage Picturesque et Historique au Brésil,* Paris 1836-39. Plate 5 and 6 both in the public domain. Plate 7 from *Travels in Brazil* by Henry Koster, 1817. Plate 8, 9, 15, 16 by the author. Plate 11 by George Huebner, c 1890. Plates 4 and 13 from Monica Lee whose book *300 Year Journaey, Leicester Naturalist Henry Walter Bates, FRS and his family 1665-1985* was privately published in 1985. Plate 17 from: Contributions to an insect fauna of the Amazon Valley. Lepidoptera-Helconiinae, *Journal of the Proceedings of the Linnaen Society* of London, read 21st November 1861. Plates 22, 23, and 24 research by Francilela Jatene Cavalcante da Silva Coordenaçao de Informaçao e Documentaçao, Museu Paraense Emilio Goeldi Belém, Brasil.

The section of 8 plates is the work of Johann Moritz Rugendas (1802 - 1858) the German artist, known for his depiction of landscapes and ethnographic subjects in Brazil in the first half of the 19th century. In his work, *Malerische Reise in Brasilien* (1827-1835), the plates are presented in four divisions [parts]. Those used in this book have been carefully chosen to depict likenesses to the Amazonia Bates would have known. Similar places can still be found in Amazonia far inland from the rivers.

Plate i is from the 1st Div, illustration number 8, **Plate ii** is from the 1st Div, illustration number 3, **Plate iii** is from the 3rd Div, illustration number 23, **Plate iv** is from the 3rd Div, illustration number 30, **Plate v** is from the 1st Div, illustration number 16, **Plate vi** is from the 1st Div, illustration number 26, **Plate vii** is from the 3rd Div, illustration number 1, **and Plate viii** is from the 4th Div, illustration number 19. The plate titles in this book are translated from the originals in *Malerische Reise in Brasilien."*

Every effort has been made to trace and acknowledge ownership of copyright material used in this volume.

Introduction

Henry Walter Bates (1825 – 1892) was the eldest son of a stocking maker in Leicester. Bates left formal school at the age of 13 to be apprenticed to Alderman Gregory, a colleague of his father, in order to learn the stocking trade and eventually take over the family business. Always eager for knowledge Bates attended the local Mechanics' Institute where he continued his education during the evenings under the tutelage of a number of inspirational teachers. In addition to this, he still found time to indulge in his hobby of beetle collecting and soon became an expert in identifying these creatures. In 1844, he met Alfred Russel Wallace, (1823 – 1913), who was for a time a tutor in English and drawing at the Collegiate School. Wallace was interested in plants but Bates soon weaned him on to beetles and the two of them became good friends collecting their specimens in the countryside surrounding Leicester. Their ever-growing interest led, three years later, to Wallace suggesting the two of them travel to the tropics to collect natural history specimens and data that might shed light on the evolution of species.

In May 1848, Bates and Wallace arrived in Belém, near the mouth of the river Amazon. At first they collected together, sending their specimens to an agent in London for disposal, but quite soon they split up, Bates to spend more than eleven years alone on the Amazons and Wallace four, although Wallace later travelled extensively in the Malay Archipelago. By the time Bates returned to England he had collected more than fourteen thousand specimens, eight thousand of which were entirely new to science. Bates spent the next few years sorting, classifying, and describing his specimens.

While collecting on the Amazons Bates had noted startling similarities between certain butterfly species, a phenomenon later ascribed as Batesian mimicry. Bates attributed this occurrence to natural selection, since palatable butterflies closely resembled non-palatable ones that predators recognised and left alone, thereby affording the palatable butterflies protection. Eventually Bates wrote a scientific paper on this providing supportive evidence for the evolutionary theories of Charles Darwin. Mimicry also influenced significant medical developments in the 20th century.

Darwin persuaded Bates to write a book about his travels in Brazil that resulted in, *The Naturalist on the River Amazons: A Record of Adventures, Habits of Animals, Sketches of Brazilian and Indian Life, and Aspects of Nature under the Equator, during Eleven Years of Travel.* 2 vols (London: John Murray, 1863). Darwin even undertook the proof reading of the early chapters proclaiming the book the best ever on natural history travel. It remains in print to this day. Bates was to write many more scientific papers but never another book declaring he would rather spend another eleven years on the great river than do so.

Bates was eventually appointed assistant secretary of the Royal Geographical Society, in effect its administrator, as he was the only paid executive in the society. He remained there from 1864 until his death from bronchitis in 1892. During this time, Bates met and dealt with the needs of almost every explorer of note during the mid Victorian age including names like Livingstone, Stanley, and Burton. Bates also helped transform the society from an explorer-based institution to an academic one furthering the universal understanding of geography in this process and placing it on the educational curriculum of our schools. He died revered more as a geographer than the travelling naturalist of his early years.

The only book-length biography about Bates is by George Woodcock. Woodcock, who was not a biologist, gives a thin account of Bates's scientific work. His remarks about Wallace are unexceptional but he is better on Bates's early life, marriage, and travels on the Amazon. Woodcock however dismisses Bates's later life too abruptly with only scant reference to the Royal Geographical Society.

Bates is however regularly mentioned in books about Wallace. Darwin had his 'moon' in Wallace, who almost pre-empted Darwin's publication of *On The Origin of Species* and Wallace, in his turn, owes some of his fame to Bates.

It is therefore interesting and revealing to look again at Henry Walter Bates, this extraordinary and talented man, and to reassess his significance and importance.

I went to Bates's Amazons in 2003 to follow in his footsteps and to explore one of the great and enduring conundrums of nature: how and why does one butterfly look like another entirely different species in order to escape predation. I travelled with my 38-year-old son David, in 2003 from Belém to Manaus, to try to decipher how species evolve in nature and are maintained. As an amateur entomologist, this lonely terrain seduced me into returning in 2005, this time travelling from Manaus to Ega (Tefé) and Fonté Boa and São Paulo de Olivença. For me, what was important was to bear witness to this place that is like no other and to engage in the process of understanding science.

How privileged I was though to experience this place at this moment in my life, almost my 70th year. Had I travelled to the Amazons in Bates's time, the vistas and ways of life would have been the same, though the journey would have undoubtedly been fraught with even more danger than I encountered. Just as I did in 2003 and 2005, I would have revelled in the dumbfounding diversity of life forms. One might imagine that it is a simple goal to understand the way things are assembled on this planet, the rules, and the order. Yet to witness the patterns of species abundance and variety with the slow realisation that this diversity was generated not by a benign and constant environment but by disturbance and inconstancy has been one of the great intellectual revelations of my life.

Special Notes

The scientific names of flora and fauna used in quotations are contemporaneous with H W Bates although in some places these have been updated for better modern understanding.

Quotations without reference are taken from Edward Clodd's commemorative edition of Bates's *The Naturalist on the River Amazons: A record of Adventures, Habits of Animals, Sketches of Brazilian and Indian Life, and Aspects of Nature under the Equator, during Eleven Years Of Travel.* (London: John Murray, 1892)

In 1848 Belém was known as Pará, Manaus as Barra (do Rio Negro), and Tefé as Ega.

Acknowledgements

My thanks are due to Laura Forsyth and Pedro Rodrigues of Journey Latin America for arranging my itinerary with skill, to Maria Gonsalves of Georgetown, Guiana, for her help in orientation to South America, to Suria Ramnarain for arrangements made in Guiana to visit Kaieteur Falls, to Rena Lentze, Guia de Turismo/Interpreters in Belém, Brazil and to Francileila Jatene C Silva of the Museu Paraense Emílio Goeldi for showing me the Bates material in the library in the Departamento de Documentação in Belém and for tracking down the official records of the award of the Order of the Rose to Bates. To Claudio and Circe Dias who invited David and me to their home on the Rio Camará on the Ilha de Marajó as house guests. To the men who caught the Anaconda (*Eunectes murinus*). To Comandante Altair of the NM Santarém for taking us to Santarém. Monte Alegre, Óbidos, and Manaus. To the forest people who gave us a glimpse of how they lived, to Ceveleno the rubber tapper on the Rio Puraquequara for his cold beer, to many others who remained in the background but nevertheless made our journey much easier by their skill and dedication.

To Helcio Amaral, Historian in Santarém for explaining the history of the town, to Jairo Oliveira, guide in Santarém for his care and attention, to Paulo Azevedo, biologist in Belém for showing me a selective butterfly breeding programme, to Julio Freitas, academic guide in Belém, for helping me find my way through the official archives, to Aldemir Junior, guide in Belém for showing me that city and the European burial places, to Alan Coelho, Pará Municipality, for help with documents, to William Overal, Professor of Entomology at the Museu Paraense Emílio Goeldi in Belém for his conversation on mimicry and Henry Walter Bates, to Igor Seligman, and André Cardoso at the Museu Paraense Emílio Goeldi in Belém for showing me their butterfly research programme, to Dr Everado Martins, butterfly breeder in Santarém, for showing me his butterfly breeding and education programme at Alter do Chão.

To Instituto Brasileiro do Meio Ambiente e Recursos Naturais Renováveis (IBAMA), for arranging our visit to the Floresta Nacional do Tapajós, and providing excellent guides for this.

To Dom Sérgio Eduardo Castriani, Bishop of Tefé and Fransisco Andrade de Lema for assisting me in trying to trace Oria, to Carlos Nader for generally helping in Tefé, to Robin and all the staff at the Reserva Ecológica Mamirauá on the Paraná do Aranapu who made our visit there so interesting and showed us the English monkey and to the Ribeirinho River people who let us into their homes.

In addition, my thanks are due to Anne Freitag, Conservatrice at the Musée cantonal de Zoologie in the Palais de Rumaine, Lausanne, Switzerland, for giving me access to Vladimir Nabokov's butterfly collection. To Anne Rowe for reading and commenting on the first draft.

To the staff of the Linnean Society, The Natural History Museum, The British Library, to Professor John Clarke and Professor Jane Ridley both of the University of Buckingham, and the many unnamed who helped me with the preparation of this book.

To Onni Crawforth, aged eight at the time, for suggesting the title for the book, and finally to Wendy Ann without whom this task could not have been completed.

I am particularly indebted to Professor Clyde Binfield, former Head of the Department of History at the University of Sheffield, and to Professor Geoffrey Alderman, Professor of Politics and Contemporary History at the University of Buckingham for their help and encouragement. My thanks also go to Dr.William Leslie Overal of the Museu Paraense Emílio Goeldi, Belém, Brazil, for his help during the preparation of this book, and to Dr.Nelson Papavero, Brazil's foremost Wallace scholar, formerly of the Museu de Zoologia, Universidade de São Paulo, Brazil for his help.

Preface

In January 1500 Vicente Yáñez Pinzón and his crew disembarked on the island of Marajó, "where they saw a new and almost monstrous animal". Thus Montalbodo described in his *Paesi nouamente retrouati* (1507) the first Amazonian animal captured by Europeans: an opossum. One of those animals together with its young was taken to Spain to be exhibited to the Royal Court, but died on the voyage. However they were exhibited and seen by many people. These creatures aroused great curiosity in the Old World about the strange and fascinating fauna of the New World.

From the 16th to the 18th century many travellers to Amazonia wrote about the flora and fauna, of which, many emanated from Portugal. Most of these writings, however, remained unpublished. A politics of secrecy was maintained by Portugal, which at the time forbade revelation of any kind about her colonies, under pain of imprisonment.

The only scientific expedition to the Amazon, sponsored by the Portuguese Crown, was conducted by Alexandre Rodrigues Ferreira between 1783 to 1792; the "Viagem philosophica pelas Capitanias do Grão Pará, Rio Negro, Mato Grosso e Cuyabá", and it gathered a rich collection. The artists accompanying the expedition illustrated many animals, plants, landscapes and Indians, but no publication resulted from this wealth of information. Eventually the material was confiscated by Geoffroy Saint Hilaire, during the French occupation of Portugal, and the specimens were taken to the Paris Museum. This allowed French naturalists the opportunity to describe many new species based on that booty.

By the end of the 18th century, Johann Centurius, Count von Hoffmannsegg, who had been travelling in Portugal to collect and study plants, managed to obtain the esteem and friendship of the King himself. He obtained the monarch's permission, an unusual achievement, to send his friend and bird collector, Friedrich Wilhelm Sieber, to Brazil in 1801 where he remained for 12 years. Sieber's journeys, travelling up the Amazon and Rio Negro, more or less mirrored those made some years later by Henry Walter Bates.His collections were eventually deposited in the Zoological Museum of Berlin University where they were studied by Dr. Carl Illiger who named those which he believed new to science.

Because of the French invasion of Portugal the Portuguese King and his Court, protected by the British navy, fled to Brazil. Once established there and having forged an alliance with the Brazilians, Dom João VI opened the ports to friendly nations including the British. In 1817, a marriage between

Pedro I and the Archduchess Leopoldina Carolina Josepha of Austria was arranged and her retinue included several scientists whose task was to describe the flora and fauna of Brazil. From this coterie, Karl Friedrich Philip von Martius and Johann Baptist von Spix collected in Amazonia from 1819 to 1820 as did Johann Natterer, who excelled as a bird collector, and prepared, on average, 2 bird skins a day for every day during his stay in Brazil.

From that time on, entry to the Amazon was officially encouraged and many naturalists travelled to the region. Foremost among them was Henry Walter Bates, an incomparable insect collector who became a distinguished natural philosopher, and made an extremely important contribution to the theory of Evolution with his discovery of mimicry.

The life and work of Henry Walter Bates, who wrote one of the most delightful books about Brazilian Amazonia, are explored in this book by Anthony Crawforth. The author has succeeded in retracing Bates's footsteps in the Amazon and presents much that is new about the man; this book is a most welcome addition to the history of evolutionary science.

Bates went into the Brazilian Amazon and returned to give Darwin the most elegant proof of feral natural selection with his insight into the mimetic colours and patterns of butterflies. It is most fitting that we recognize and reacquaint ourselves with this early Darwinian, so admirably treated in this fine book. Both Bates and Crawforth inspire us to follow them into the tropical forest where, without a doubt, new scientific discoveries remain to be found.

Dr. William Leslie Overal, Museu Paraense Emílio Goeldi, Ministério de Ciência e Tecnologia, Belém, PA, Brazil, and Dr. Nelson Papavero, Museu de Zoologia, Universidade de São Paulo, São Paulo, SP, Brazil.

March 2009.

Part one

Early Development

Chapter 1

The Evolution of an Evolutionary Man

A number of early eighteenth and nineteenth century travellers to Brazil influenced Henry Walter Bates and his friend, Alfred Russel Wallace, in their search for an understanding of the origins of species. These prominent eighteenth and nineteenth century travellers were intelligent and versatile men, of varied origins, who had the opportunity to demonstrate unusual energy and resourcefulness. One of them, Charles Darwin, needs little introduction, having been continuously and meticulously analysed in the many books that examine his life and work. Wallace, on the other hand, sidelined for many years as a mere shadow of Darwin, has recently been re-evaluated and there is now a greater understanding of his role as co-discoverer with Darwin of the origins of species. This has resulted in the publication of a run of books placing Wallace near the pinnacle in this field of discovery.[1]

Bates however remains relatively unknown, hardly more than a character figure in the account of the heroic duo, Darwin and Wallace. Nevertheless, a theory about mimetic resemblances (mimicry) attaches to Bates's name and for this he is remembered though still not widely celebrated. Later Wallace, who began his travels with Bates on the Amazons, had similar thoughts about mimetic resemblances in the Malay Archipelago. Today the theory of mimicry attracts both believers and disbelievers, but the fact remains that had Bates not first defined his idea, Darwin and Wallace would probably have been denied access to a very specific process of thought. The importance of this theory of mimicry is that it led to a greater understanding of evolution. Therefore, this book sets out to bring Bates out from the shadows, and place this extraordinary man in a more elevated scientific

---

[1] These books are:

Knapp, Sandy, Footsteps in the Forest, Alfred Russel Wallace in the Amazon, (London: The Natural History Museum, 1999).

Raby, Peter, *Alfred Russel Wallace, A Life*, (London: Chatto & Windus, 2001).

Berry, Andrew, Infinite *Tropics, An Alfred Russel Wallace Anthology*. (New York: Verso, 2002).

Finchman, Martin, *An Elusive Victorian: The Evolution of Alfred Russel Wallace*. (Chicago and London: The University of Chicago Press, 2004).

Slotten, Ross A, *The Heretic in Darwin's Court*, (New York: Columbia University Press, 2004).

context. If Darwin needed Henslow, then Darwin and Wallace needed Bates. [2]

In this story, we observe Bates in the intimacy of his family, as the young boy searching for knowledge and later as the untiring explorer and adventurer. We analyse him as a scientific theorist and as an itinerant collector, jotting down with meticulous care the daily detail of his collecting. In later life, we see him wrestling with the problem of acceptance by the scientific elite of his day despite the fact that in the nineteenth century, most people were assigned predictable social roles and status and were thus expected to identify with the standard models that were associated with gender, social class, nationality or religion.

We also focus on the Amazon, with a retelling of Bates's scientific journey as I experienced it following in his footsteps. Sometimes using his own words, supplemented by my own observations, I attempt to effectively interpret those events in the Amazons for the modern reader. This should enable the person who reads Bates's book, *The Naturalist on the River Amazons: A Record of*

[2] Loewenberg, Bert James, Darwin, Wallace, and the Theory of Natural Selection including the Linnean Society Papers. (Cambridge, Massachusetts: Arlington Books, 1959) 12.
In addition, as shown in this letter in the Stecher Collection in the Cleveland Health Sciences Library, from Darwin to Bates, dated 22nd November 1860, Darwin recognised Bates's anticipation of evolutionary theory soon after Bates's return from the Amazons. It reads:

My dear Sir

I thank you sincerely for writing to me & for your very interesting letter. Your name has for very long been familiar to me, & I have heard of your zealous exertions in the cause of Natural History. But I did not know that you had worked with high philosophical questions before your mind. I have an old belief that a good observer really means a good theorist & I fully expect to find your observations most valuable. I am very sorry to hear that your health is shattered; but I trust under a healthy climate it may be restored. I can sympathise with you fully on this score, for I have had bad health for many years & fear I shall ever remain a confirmed invalid.

I am delighted to hear that you, with all your large practical knowledge of Nat. History anticipated me in many respects & concur with me. As you say I have been thoroughly well attacked & reviled (especially by entomologists, Westwood, Wollaston & A. Murray have all reviewed & sneered at me to their hearts' content) but I care nothing about their attacks; several really good judges go a long way with me, & I observe that all those who go some little way tend to go somewhat further. What a fine philosophical mind your friend, Mr Wallace has, & he has acted in relation to me, like a true man with a noble spirit. I see by your letter that you have grappled with several of the most difficult problems, as it seems to me, in natural History, such as the distinctions between the different kinds of varieties, representative species &c.

Perhaps I shall find some facts in your paper on intermediate varieties in intermediate regions, on which subject I have found remarkable little information. I cannot tell you how glad I am to hear that you have attended to the curious point of Equatorial refrigeration. I quite agree that it must have been small; yet the more I go into that question the more convinced I feel that there was during the glacial period some migration from N. to S. The sketch in the Origin gives a very meagre account of my fuller M.S. Essay on this subject.

I shall be particularly obliged for a copy of your paper when published; & if any suggestions occur to me (not that you require any) or questions I will write & ask.

Pray believe me, with respect & good wishes, My dear Sir, Yours sincerely, C. Darwin

Ps. I have at once to prepare a new Edit of the Origin, & I will do myself the pleasure of sending you a copy; but it will be only very slightly altered. Cases of neuter ants, divided into castes, with intermediate gradations. (which I imagine are rare) interest me much. V.Origin on the Driver Ants p. 241. (please look at the passage).

Adventures, Habits of Animals, Sketches of Brazilian and Indian Life, and Aspects of Nature under the Equator, during Eleven Years of Travel. (hereafter referred to as *The Naturalist on the River Amazons*), to place Bates and his work in a proper historical perspective, for without discussing advances made since he wrote his book, it is impossible to appreciate his achievements.[3]

In many books and articles, Bates is relegated to a footnote or its equivalent, and denied his full measure of greatness on two grounds. One was that he was always in the shadow of greater men and his work was deemed to be less important and therefore the glory should belong to them rather than to him. The other was that his views have been so completely superseded that he cannot now deserve any exceptional praise or a prominent place in the annals of science.

This latter point is the easier of the two to disprove, for it is untrue to say that his views have been superseded. The only sense in which this word could be used in relation to his scientific theories would be if they have been disproved and dismissed, or if they lay outside the main line of scientific relevance, neither of which is the case. The chief sense in which his ideas may have been superseded, if that term is to be used, is a wholly complimentary one, namely that his theory of mimicry met with such universal acceptance that it is now taken for granted among biologists, much as the particulate structure of matter is taken for granted among physicists and chemists. Thus, it has become part of the general background of biological thought, even though the original idea has evolved since Bates's time.

The first point is clarified as the book develops, in that Bates is shown as equal to Wallace as a contributor to a more detailed understanding of general biology and because of his extraordinary mastery of entomology, taxonomy and the collector's technique.

Bates's destiny was to become a hosier in the family business. This could have become a life of tedium in Leicester, when by chance he met another enthusiastic young naturalist named Alfred Russel Wallace at the town's public library. Wallace had just been appointed as a teacher at the local Collegiate School and was particularly interested in botany. Bates had already

---

[3] Bates, Henry Walter, *The Naturalist on the River Amazons: A Record of Adventures, Habits of Animals, Sketches of Brazilian and Indian Life, and Aspects of Nature under the Equator, during Eleven Years of Travel.* 2 vols (London: John Murray, 1863) This book's reputation suffered from the abridged second edition of 1864 and all succeeding editions except when Clodd, 1892, persuaded Murray to reprint the original text with his memoir of Bates. The popularised editions removed the science of the original and rearranged Bates's journeys for easy reading. They also sanitised Bates achievements and made it difficult for the reader to understand Bates's incredible journey and distinguished service to both science and geography. Today only the revised edition remains in print.

published some short notes about beetles in volume 1 of *The Zoologist* [4] and was an enthusiastic amateur collector. Bates broadened Wallace's interests to include entomology and, when Wallace left Leicester, he kept in touch with Bates. Two years later, Wallace suggested they should travel together to the Amazon as peripatetic collectors, and while Wallace stayed on the Amazon for only four years before moving on to the Malay Archipelago, Bates remained there for more than eleven years. On returning to London, Bates began sorting his collections and publishing important entomological papers culminating in 1862 with *'Contributions to an Insect Fauna of the Amazon Valley. Lepidoptera: Heliconiidae'* [5] which included a geographical analysis of the distribution of species and announced the theory of protective resemblances or mimicry. In 1863, with encouragement from Charles Darwin he published his book *The Naturalist on the River Amazons,* [6] which quickly reached classic status in the field of literary scientific travel writing. The rest of his life seemed to have been something of an anticlimax, devoid of further physical exploration or overseas adventure. His later work as assistant secretary to the Royal Geographical Society, beginning in 1864, was extremely demanding, though it still allowed him opportunities to assist many other naturalists and travellers in a variety of ways. Although he was to continue to produce good descriptive entomological work, he never travelled or ventured into the realms of scientific philosophical theory again. However, as a taxonomist he was unequalled.

Renouncing the Regnant Faith of Immutability

In the late 1840s one or other of two beliefs were still strongly held by most Victorians. The first was that the earth had been created within historical times. In 1650, James Ussher, the Archbishop of Armagh, had pinpointed the moment of creation as the evening before Sunday 23rd October 4004 B.C. If the world had been created thus, then the multitudes of living things had surely been created in the same way, by a stroke of

---

[4] Bates, H W, Notes on Coleopterous Insects frequenting damp places, *The Zoologist,* 1843, vol. 1, 115.

[5] Bates, H W, Contributions to an insect fauna of the Amazon valley. - Lepidoptera: Heliconiinae. *Journal of the Proceedings of the Linnean Society of London 6,* (1862) 73-77.

[6] Letter, Darwin to Bates, Down, Bromley, Kent, 4th April 1860, *'Have you ever thought of publishing your travels?' The Naturalist on the River Amazons* (Bates 1863) was announced in John Murray's list of new works for April 1863 (*Publishers' Circular* 26 (1863): 172, 193). Darwin had encouraged Bates to publish with Murray, and had written to the latter supporting Bates. See Darwin *Correspondence* vol. 9, letter to H. W. Bates, 25th September (1861), and *Correspondence* vol. 10, letter to John Murray, 28th January (1862).

omnipotent genius that produced men, fish, snakes, birds, and the micro insects, not exactly in the seven days and nights of the Bible but at least in some comparable period. The alternative belief was that even if the seven days and seven nights were purely allegorical, creation had still been a matter of separate events. Bates and Wallace simply rejected these ideas as all the evidence they could see for themselves suggested otherwise.[7]

Archbishop Ussher's pronouncement has tended to make him a laughing stock. However Ussher's was a serious work of scholarship that began a tradition of inquiry into geochronology at Trinity College, Dublin, that led directly to the radiometric dating techniques that have now established the Earth's age at 4567 million years.

Formative Influences on Bates 1840 - 1848

Carl von Linné (1707-1778)[8]

The first definitive influence on the young Bates and Wallace was the writing of Carl von Linné (1707-1778) or Carolus Linnaeus as he is better known. Linnaeus trained in medicine, but spent the greater part of his working life teaching natural history at the University of Uppsala in Sweden. As a young man he travelled extensively in the northern parts of his country on what became a journey of considerable intellectual inspiration. Later, although travelling more widely, his first-hand experience of natural history was limited to the Holarctic region. His students, however, travelled to many parts of the world, endeavouring to apply his teachings in a variety of situations. One of them, Daniel Solander (1733-1782), accompanied Captain Cook on his first voyage, and together with Joseph Banks (1743-1820) made the first botanical observations of Australia in the year 1770.

Linnaeus's fame rests upon his work as a taxonomist of the natural world, and in particular on his system for the sexual classification of plants. He began this taxonomic encyclopedia, his celebrated *Systema Naturae* of 1735 (while still a student) and *Systema Vegetabilium* of 1774. In his work, Linnaeus was striving to find the order and pattern of nature and to present it in the form of a simple, comprehensible, and coherent system. He divided nature into three kingdoms: animal, vegetable, and mineral, which were then treated more or less separately. Each kingdom was divided successively into

---

[7] Bowler, Peter J, *Evolution, The History of an Idea:* (Berkeley: University of California Press 2001) 4, 29.

[8] Hagberg, Knut, *Carl Linnæus*, (London: Jonathan Cape, 1952).

classes, genera and species, all of which were presented in the form of a table that enabled the complete classificatory system of a particular kingdom to be taken in almost at a glance. Most, if not all, natural history writers thereafter assembled their descriptions of species in this manner.

Bates, like all aspiring young students of the natural sciences, was aware of Linnaeus's work before his journey to Brazil, and he had taught himself its fundamentals during evening classes at the Mechanics' Institute in Leicester. Later, his scientific papers on insects showed not only a grasp of this methodology but an ability to advance Linnaeus's science. Without this fundamental knowledge the foundation of Bates's taxonomic skill would have floundered early on in his travels.

From a twenty-first century point of view, there may seem to be a great many deficiencies in Linnaeus's tabular representation of nature. His work should not be judged however in modern terms but as far as it is possible for us to do so, through eighteenth-century eyes. It was Linnaeus's opinion that his approach would reveal the very design of the animal kingdom that God had employed at the time of the creation. This was the task to which his life's work was directed: the revelation of the order of God's design and the appropriate naming of the products of His creation.

The importance of these ideas was that they made young naturalists like Bates and Wallace think about the design of the natural world, and eventually to address the question of whether the superabundance of the life they could see was placed there by the hand of God or arrived at by some other means. It also gave them a structure for all nature with a naming system that has endured.

Baron Friedrich Wilhelm Karl Heinrich Alexander von Humboldt (1769-1859), whose El Dorado was the flora and fauna of the Amazons.[9]

Another writer who exerted a considerable influence on Bates and Wallace was Alexander von Humboldt (1769-1859). He had been born into a Prussia where science was considered an unworthy subject for a person of his social standing, a situation that was quite the reverse of the attitude then prevailing in England. The year of his death, 1859, saw the publication of *On the Origin of Species* and the great explosion of knowledge to which Humboldt had contributed so much.

[9] Botting, Douglas, *Humboldt and the Cosmos,* (London: Michael Joseph 1973) 75.

Few scientists had the opportunity to visit the South American colonies before the end of the eighteenth century as the Spaniards and the Portuguese kept the door to their territories firmly shut. However, Humboldt received a permit issued by the King of Spain to explore the Spanish realms, not just because of his reputation as a scientific thinker of his day but also in view of the fact that he was by trade a geologist and mining engineer. The King was hoping for news of valuable mineral deposits that could augment the royal coffers. To the Spanish, all of South America was still a land of hidden golden treasure awaiting discovery, so Humboldt was directed to investigate, from a detached and enlightened point of view, the legend of El Dorado.

Once admitted to these hidden places in 1799, Humboldt and his companion Aimé Bonpland were the first Europeans with both the intelligence and scientific knowledge to interpret fully the extensive opportunities this exploration opened up. They were astonished by what they found and as Humboldt wrote to his brother shortly after landing in South America:

> "What a fabulous and extravagant country we are in. What trees, coconut palms, 50 to 60 feet high; *Poinciana pulcherrima* with a big bouquet of wonderful crimson flowers; ... a whole host of trees with enormous leaves and sweet smelling flowers as big as your hand, all utterly new to us. As for the colours of the birds and fishes-even the crabs are sky-blue and yellow! Up till now we've been running around like a couple of mad things; for the first three days we couldn't settle to anything; we'd find one thing, only to abandon it for the next. Bonpland keeps telling me he'll go out of his mind if the wonders don't cease soon." [10]

Humboldt and Bonpland discovered more than three thousand plant species new to science and later Humboldt's thirty volumes were published and translated into English to describe their botanical, entomological, zoological, geological, astronomical, meteorological, historical, and artistic observations and discoveries. The formidable mass of scientific data collected during this five-year expedition laid the foundations for modern physical geography, and Humboldt's investigations as a naturalist established the whole concept of plant geography.[11]

---

[10] Botting, Douglas, *Humboldt and the Cosmos,* (London: Michael Joseph 1973) 76.
[11] Botting, Douglas, *Humboldt and the Cosmos,* (London: Michael Joseph 1973) 155.

In his books, Humboldt demonstrated that South America was strange and fantastic and more rich and exotic than the most inspired artists or writers of the age had ever dreamed. The volumes published under his personal supervision contain the first accurate, illustrated impression of the landscape, its people, flora, and fauna.

Plate 3.
Return of the Naturalist's black assistant

After this expedition, South America gradually became more easily accessible to Europeans and following the Napoleonic wars, several French artists went to Rio de Janeiro to teach.

Humboldt had not been able to take an artist on his expedition so artists illustrated his books working from his own rather rudimentary sketches and descriptions, together with dried specimens of plants, dead animals and birds. Nevertheless, no one was more aware than Humboldt of the need for painters to represent nature as it was in its natural state. In his least specialised and therefore probably most widely read book *Views of Nature* (1808),[12] he remarked that it would be worth a great artist's while to study the various species of plants and nature generally, not in hothouses or from descriptions by naturalists, but in the '*grand theatre of tropical nature*', believing such work would be of service not merely to the arts but also to science. Half a century later he was able to say that what Hodges, the artist on

[12] Humboldt, Alexander von, *Views of Nature*, translated by E Otté and H Bohn, (London: 1880)

Captain Cook's first expedition, had done for the islands of the South Seas, had been accomplished for tropical America with much more style and with greater accuracy by the artist Johann Moritz Rugendas (1802-1858).[13]

Humboldt returned to Europe to devote almost thirty years and his considerable fortune to the publication of his books about South America and then, in his sixties, set off on a further expedition to Siberia. In the last years of his life, Humboldt strove to complete his greatest work - *Cosmos* - his vision of the nature of the world.[14]

Humboldt is important because not only did he describe tropical nature in detail, but he also measured it precisely, using more than seventy of the best scientific instruments his private wealth could buy. To his empirical descriptions of the physical environment, he matched an intensely aesthetic approach to nature, creating a view of the tropics as a sublime place. Scientist, explorer, and diplomat, Humboldt left his name on the maps of five continents; over a thousand places in the world are named after him and a crater on the moon bears his name. In private he was emotional, unflinchingly hardworking, homosexual, gentle and in later years, pathetically poor. His influence on Bates and Wallace cannot be overestimated.

Johann Moritz Rugendas (1802-1858)[15]

The artist Johann Moritz Rugendas, first noticed by Humboldt in c1819, went to South America in 1821 at the age of nineteen as draftsman on an expedition that the Russian Consul-General, Baron von Langsdorff (1774-1861), planned to lead to the interior of Brazil. Rugendas stayed for four years, preparing a large number of drawings with which he returned to Europe. Wanting to publish them, he contacted Humboldt who warmly responded to his work and commissioned some drawings of plants. Lithographs of a hundred of his Brazilian views were published with a brief text translated into French as *Voyage pittoresque dans le Brésil* (issued in four parts, Paris 1827-35). His spontaneously bold sketches are tantalisingly

[13] Some illustrations in this book are taken from Malerische Reise in Brasilien. Paris, Engelmann & Cie., 1835 or Brazilian Topography by Johann Moritz Rugendas an important work on nineteenth century Brazil. Rugendas accompanied Baron Langsdorf's 1825 scientific expedition to Brazil as the official artist, but abandoned the party in Rio de Janeiro and ventured out alone. He focused on the customs of Brazilian natives and imported slaves, but also recorded interactions between white settlers and the indigenous population. They are a significant iconographic record of two important exploratory expeditions and are more or less contemporary with Bates's and Wallace's travels and a unique view of the Amazons, as they would have been seen it. Published by permission of Massimo de Martini, Altea Gallery, 35, St George Street, London.

[14] Humboldt, Alexander von, *Cosmos*, translated by E Otté, 5 vols (London: 1848-1858).

[15] Rugendas, Johann Moritz, *Voyage pittoresque au Brazil*, Paris:1835 (1930s Sao Paulo reprint).

seductive to the viewer and drawn as if he barely had time to record the scenes before passing impatiently on to the next one.

Rugendas rejected the old conventions, paid great attention to nature, and was keenly sensitive to the peculiarities of light in the tropics. Rugendas's reputation was for the uncommon graphic quality found in his work accurately representing tropical nature and savage Indians. He sought, above all, to convey an impression of the emotions South America had aroused in him. Engravings by Rugendas have been chosen to illustrate this book, as they are more or less contemporary with Bates and show the forest and its people, as he would have encountered them.

Carl Friedrich Philipp von Martius (1794-1868) and Johann Baptist von Spix (1781-1826)[16]

Two Bavarian naturalist explorers, Carl Friedrich Philipp von Martius and Johann Baptist von Spix did for the Amazon what Humboldt had done for the Orinoco. We know Bates and Wallace admired them as in their early days in Belém they had looked for and found the chácara (small rural property) Martius and Spix had occupied. Even though it was then overgrown and dilapidated, Bates commented on the significance of this event. [17]

The son of a Bavarian doctor, Spix was born in 1781 and his early academic career, like Darwin's, included studies in theology, medicine, and the natural sciences. He qualified as a medical doctor in 1806. In 1808, he was awarded a scholarship by King Maximilian Joseph of Bavaria to study zoology in Paris, where he mixed with the leading biologists and naturalists of the time, including the French biologist Jean-Baptiste Lamark. In Paris, Spix's reputation as a scientist grew, and in October 1810, the King acknowledged this with an appointment to the Bavarian Royal Academy of Sciences as the curator of its natural history exhibits.

---

[16] From 1823 to 1850 Martius published Historia Naturalis Palmarum in three immense volumes. He set the outlines of the modern classification of palms and prepared the first maps of palm geography. He described the palms of Brazil in the second volume and, in the third, those of the rest of the world, as far as he was able to draw on the researches of others. Historia Naturalis Palmarum was reprinted in two volumes in 1971. Spix published Geschichte und Beurtheilung aller Systeme in der Zoologie nach ihrer Entwicklungsfolge von Aristoteles bis auf die gegenwärtige Zeit. 1811 - Nürnberg, Schrag'sche Buchhandlung I-XIV;710pp. In addition to the 4-volume narrative of the expedition, Reise in Brasilien in den Jahren 1817 bis 1820 (Munich,1823-1831).

[17] Clodd, Edward, Memoir in unabridged (1863) commemorative Edition of The Naturalist on the River Amazons: A record of Adventures, Habits of Animals, Sketches of Brazilian and Indian Life, and Aspects Nature under the Equator, during Eleven Years Of Travel. (London: John Murray, 1892) 111.

In 1817, the two scientists found themselves on board Brazil-bound ships in the suite of Archduchess Leopoldine Caroline as members of an expedition under the patronage of the Emperor of Austria.[18]

Martius, whose interests were mainly botanical with a passion for palms, was also a gifted academic, thirteen years younger than Spix. Spix was to concentrate on animals, the local people, and geological recording, including the collection of fossils. Martius was to devote his energies to botanical investigation, including soil types and the study of how plants spread to new lands. As a bird collector himself, the King hoped that the two men would bring him novel and matchless prizes for his collection from their expedition to the New World.

Martius and Spix's superbly illustrated books brought the fetid gloom of the jungle, with its immensity and its dangers, into mild-mannered and peaceful nineteenth century drawing rooms, and conveyed adventure by proxy to the reader with powerful and disturbing images of a vast, stifling, hostile continent of impenetrable and dangerous forests.

The New Philosophy

Robert Chambers (1802-1871)[19]

The Edinburgh publisher, Robert Chambers (1802-1871), was to influence both Bates and Wallace as the only writer in Britain before Darwin to put forward a detailed argument for biological evolution. Though there were several men of minor reputation who proposed versions of natural selection in rather obscure publications, these little-known writers had virtually no influence on public opinion. By contrast in 1844, the anonymously published work of Robert Chambers entitled *Vestiges of the Natural History of Creation (Vestiges)* caused something of a sensation. Chambers's work was not based upon the theory of evolution by natural selection, but rather on the cosmic evolutionism of the German Natural Philosophers. Although the doctrines of *Vestiges* never achieved any degree of popular acceptance, the work fulfilled a considerable role in preparing the public psyche for the

---

[18] The Emperor's daughter; (Archduchess Leopoldine Caroline) was to marry the son of John VI, King of Portugal. King John had been forced to live in Brazil (since 1808) following the invasion of his homeland by Napoleon Bonaparte of France. (John VI's son eventually became Dom Pedro I, Emperor of Brazil).

[19] Chambers, Robert, *Vestiges of the Natural History of Creation*, (London: 1844), Also referred to by Wallace, *My Life*, vol 1, (London: Chapman and Hall, 1905) 254-257.

eventual acceptance of Darwin's doctrines. Both Bates and Wallace were influenced by the revolutionary ideas contained in the book.

Thomas Robert Malthus (1766 – 1834)[20]

Malthus put forward the theory that unrestrained population growth always exceeded the means of subsistence. Controlled population growth on the other hand could be kept in line with food supply by positive checks like starvation, disease, and death. Malthus's proposition implied that population always had a tendency to grow beyond the food supply. As a result, he argued, any attempt to upgrade the condition of the lower classes by increasing their incomes or improving agricultural productivity would be fruitless, as the extra means of subsistence would be completely absorbed by an increase in population.

As long as that tendency remained, Malthus argued that the perfect society would always be out of reach. He reduced this to a simple formula. Populations tend to increase geometrically, (i.e. 2-4-8-16-32-64-etc), whilst food supplies increase arithmetically (i.e. 2-4-6-8-10-12-etc). The result is demand always exceeds supply and populations are ultimately controlled by starvation. In his much expanded and revised 1803 version of the essay, Malthus concentrated on bringing empirical evidence to bear on the issue, much of it acquired on his extensive travels to Germany, Russia and Scandinavia. He also introduced the possibility of moral restraint or voluntary abstinence for bringing the unchecked population growth rate down to an acceptable level, thus balancing supply and demand. In practical terms, this meant inculcating the lower classes with middle-class virtues. He believed this could be achieved with the introduction of universal suffrage, state-run education for the poor and more controversially, the elimination of the Poor Laws and the establishment of an unfettered nation-wide labour market. He also argued that once the poor had a taste for luxury they would demand a higher standard of living for themselves, and voluntarily reduce the birth rate. The essay transformed Malthus into an intellectual celebrity but he was reviled by many as a hardhearted monster, a prophet of doom, and an enemy of the working class. The ridicule and invective rained down on him by the chattering and pamphleteering classes was relentless. Nevertheless, a sufficient number of people recognised his essay as the first

[20] Malthus, Thomas Robert, *Population, The First Essay*, (Michigan: University of Michigan Press, 1959) (Introduction to the essay).

serious economic study of the welfare of the lower classes. In 1805, Malthus was appointed Professor of Modern History and Political Economy at the East India College in Haileybury, becoming England's first academic economist. His influence on Bates, Wallace, and Darwin was considerable because it was his ideas that led them to think about the survival of the fittest.

Charles Waterton (1782 -1865)[21]

Always in the shadows but of considerable importance is Charles Waterton (1782-1865), Squire of Walton Hall (27th Lord of Walton) eccentric, traveller, naturalist and conservationist. This pioneering naturalist opened the world's first nature reserve in the grounds of his estate at Walton Hall near Wakefield in the West Riding of Yorkshire. His family, originally from Lincolnshire, had migrated to Yorkshire several centuries before and many of its pre-Reformation members were eminent in the service of the State. Staunch Royalists as well as Catholics, they were persecuted for their belief, and became impoverished.

Receiving a Catholic education at Stonyhurst, Waterton became a good Latin scholar and developed a passion for natural history, particularly ornithology. On leaving school, he travelled first to Spain and was in Malaga when the great plague occurred there. He returned home in poor health and went in search of an even warmer climate undertaking the management of his uncle's estates in Guiana. He lived in Georgetown from 1804 to 1812 but made occasional visits home. In 1806 his father died, leaving him heir to Walton Hall. After handing over the Guiana estates to local management he set about exploring the interior of Guiana and, commencing in 1812, made the four expeditions that he described in his book, *Wanderings in South America.*

Because of his familiarity with the country, Waterton was well suited to exploring the interior of Guiana and made many very valuable additions to European knowledge of its fauna, especially bird life, and its flora. The main purpose of his first journey was to collect the deadly *wourali* poison that induces anesthesia, as it was hoped it would prove to be of medical benefit in England.

---

[21] Malthus, Thomas Robert, *Population, The First Essay*, (Michigan: University of Michigan Press, 1959) (Introduction to the essay).

Waterton's experiments with the poison proved that its deadly effects could be neutralised by artificial respiration during the period of its activity, a forerunner of modern anesthetic techniques. His other services to science included his unrivalled knowledge of the living habits of animals, which he combined with a new method of preserving skins, thus single-handedly raising taxidermy from a sorry and often ridiculous handicraft to an absolute art.

In 1865, after surviving many perils abroad, Waterton met his death at the age of 83 on his estate through an internal injury caused by stumbling over a briarroot. He was a consummate storyteller, his anacondas and cayman seemed to get larger each time he referred to them! However, Darwin, who visited him at Walton Hall, recognised the value of his work, and his friend Thackeray testifies to his moral worth in a well-known passage in *The Newcomes*: 1855, chapter XXXV:

> "I could not but feel a kindness and admiration for the good man. I know his works are made to square with his faith; that he dines on a crust, lives as chastely as a hermit, and gives his all to the poor."

At the time of its publication in 1825, his book *Wanderings in South America* was the most popular and widely read travel narrative on South America, sought after by all would-be adventurers. It was included in the libraries of the Mechanics' Institutes and in public libraries as evidenced by the labels that appear on occasional copies auctioned in book sales today. There is reason to believe Bates and Wallace both read the book, and having done so would surely have been motivated by this quite extraordinary explorer.

Sir Charles Lyell (1797-1875)[22]

Another significant influence on Bates and Wallace was Sir Charles Lyell. Wallace makes the point that both he and Bates read Lyell when he comments in a letter:

> "I was much pleased to find that you so well appreciated Lyell. I first read Darwin's 'Journal' three or four years ago, and have lately re-read it. As the journal of a scientific traveller, it is second only to

[22] Lyall, Charles, *Principles of Geology*, 3 vols, (reprint) (Chicago: University of Chicago Press, 1990).

Humboldt's 'Personal Narrative', as a work of general interest, perhaps superior to it.' My reference to Darwin's 'Journal' and to Humboldt's 'Personal Narrative' indicate, I believe, the two works to whose inspiration I owe my determination to visit the tropics as a collector'."

From his boyhood, Lyell was fond of natural history, particularly entomology, a taste that he cultivated at Bartley Lodge in the New Forest, where his family had moved soon after his birth. At that time, in the early 1800s, butterflies were plentiful in the New Forest, in some places more so than would be seen in the Amazon today. This continued to be the case until the late 1940s when a mixture of environmental change, toxic pest control agents and atmospheric pollution reduced populations to much lower numbers.

By the 1820s Lyell had already begun to plan his chief work, *The Principles of Geology*. The supplementary title, '*An Attempt to Explain the Former Changes of the Earth's Surface by Reference to Causes now in Operation*,' defines the task to which Lyell devoted his life. A journey with Sir Roderick Murchison in 1828, (later to be President of the Royal Geographical Society at the time that Bates was the assistant secretary) gave rise to joint papers on the volcanic district of Auvergne and the tertiary formations of Aix-en-Provence. After parting with Murchison, Lyell developed the idea of dividing this geological system into groups, characterised by the proportion of recent to extinct species of shells found in each group. The names he gave to the groups are now universally adopted: Eocene *(dawn of recent)*, Miocene *(less recent)*, and Pliocene *(more recent)*, and he drew up a table of shells to illustrate this classification. The first volume of the *Principles of Geology* appeared in 1830, and the second in January 1832. It was a great success and the two volumes had already reached a second edition in 1833 when the third volume, dealing with the successive formations of the earth's crust, was published. Between 1830 and 1872, eleven editions of this work were available, each enriched with new material to form a complete history of the progress of geology during that period. Only a few days before his death, Lyell finished revising the first volume of the 12th edition; his nephew Leonard Lyell completed the revision of the second volume, and the work appeared in 1876.

In August 1838, Lyell published the *Elements of Geology*, which, from being originally an expansion of one section of the *Principles*, became a standard work on geology. Bates and Wallace read this at the Mechanics' Institute in Leicester. Lyell's works explained, in simple terms, that if the rocks were millions of years old so were the relics of organic life found within them, thereby making evolution an incontrovertible fact. Lyell died on 22nd February 1875 and was buried in Westminster Abbey.

William H Edwards (1822-1909)[23]

Before finally deciding to travel to South America, Bates and Wallace read the American naturalist and butterfly collector William H Edwards's book, *A Voyage up the River Amazon, Including a Residence at Pará* that had been published in London by John Murray in 1847. Edwards was of a similar age to Bates and Wallace and the book made a great impression on them both. As Wallace was to comment:

> "This little book was so clearly and brightly written, described so well the beauty and the grandeur of tropical vegetation, and gave such a pleasing account of the people, while showing that expenses of living and of travelling were both very moderate, that Bates and myself at once agreed that this was the very place for us to go to if there was any chance of paying our expenses by the sale of our duplicate collections."[24]

Bates and Wallace were also fortunate to meet Edwards in London. He provided them with valuable letters of introduction to his friends and acquaintances on the Amazons. The book itself is of little scientific value but its influence, mainly because of its passion and flowery language, was conclusive in confirming for Bates and Wallace that the project they had in mind could be achieved.

Charles Robert Darwin (1809-1882) Humboldt's Disciple [25]

In 1839 Charles Darwin published his account of the voyage of the Beagle as volume III of *Narrative of the Surveying Voyages of his Majesties ships*

[23] Woodcock, George, *Henry Walter Bates: Naturalist of the Amazons*. (London: Faber & Faber) 25.
[24] Edwards, William H,, A Voyage on The River Amazon including a Residence in Pará, (London: John Murray, 1847).
[25] Browne, Janet. *Charles Darwin: Voyaging: Volume I of a biography:* (London: Pimlico, 1995) & Browne, *Charles Darwin: The Power of Place: Volume II of a biography*: (London: Jonathan Cape, 1995).

Adventure and Beagle, between the years 1826 and 1836. The first volume of the *narrative* contains Captain King's account of an expedition in the *Adventure* between 1826 and 1830, which surveyed the coasts of Patagonia and Tierra del Fuego. The second volume and appendix describe the second voyage of the *Beagle* under Captain Fitzroy between 1831 and 1836, which visited Brazil, Argentina, Tierra del Fuego, Chile, Peru, the Galapagos Islands, Tahiti, New Zealand, Australia and other islands and countries. The third volume is Darwin's own account of the voyage of the *Beagle.* As his first published book, it is an outstanding account of natural history exploration and describes the fieldwork ultimately leading to *On the Origin of Species.* The five years of the voyage were the most important event in Darwin's intellectual life and probably also in the history of biological science. When Darwin sailed away from England he had no formal scientific training but he returned a man of science knowing the importance of evidence. Moreover, he was almost convinced that species had not always been as they were since the creation but had undergone change. The experiences of his five years journeying built up into a process of epoch-making importance in the history of thought. The popular version of volume III was published in 1845 by John Murray and read by both Bates and Wallace. Darwin was a superb writer, able to hold the attention of any reader by the quality and style of his prose, persuasive descriptions, and later arguments.

After 1859, and following Bates's return from Brazil, Darwin's great authority as an evolutionist stemmed first from having provided what turned out to be the most persuasive account of how evolution worked and specifically, how the plants and animals of the natural world had achieved their present forms. It was the most powerful idea ever to occur to man. The mechanism, the principle of natural selection, to which his book *On the Origin of Species* was committed to proving was simple enough. Its vital message was that the huge variety of living things had resulted from the interaction of three principles.

The first was that all organisms reproduce, the second that even within a given species every organism differed to some extent from any other, and the third was that all organisms competed for survival.

If the environment changed or if new organisms entered the territory of established ones, then those individuals best adapted to the changed situation would eventually outbreed those less well adapted. In time, modification or adaptation to the environment and later geographical separation might induce enough change in the descendants of the original to create a new species.

The engine of change was called *natural selection* on the basis that nature was accomplishing what human breeders of domestic animals achieved when they bred a new kind of horse or dog by scientifically mating those individual animals that displayed the characteristics the breeders wanted to develop in the offspring.

Nature, Darwin always claimed, harboured no comparable purpose; in fact, it had no aim at all. The enormous diversity of living organisms resulted, not from a plan of intention but from the accidents of history, from changes in climate, weather, geology, and food supply, or the increase or decrease in the presence of enemies to which an animal or plant might be subjected. There was no place in the Darwinian world of natural selection for Creation or any supernatural force, and that, of course, was why many Victorians found Darwin's writings a danger to established Christianity, or any religion for that matter. There was however, a place for environment as a source of change. In Darwin's concept of natural selection, environment was the rule to which an organism must adapt or adjust and failure to do so would result in extinction. Paradoxically enough, this way of looking at Darwinian natural selection would prove to be influential in giving an environmental along with a biological cast to the thinking of social scientists.

Nowhere within the *On the Origin of Species* is evolution or natural selection applied to human beings. Only on the penultimate page of his book does Darwin become bold enough to mention human beings at all, and then it is only to suggest that *'much light will be thrown on the origin of man and his history'*. It is at this point that Bates is involved as a correspondent with Darwin after the publication of his (Bates's) *Contributions to an Insect Fauna of the Amazon, Lepidoptera: Heliconiidae* published in the Transactions of the Linnaean Society in 1862. This had a significant influence on Darwin. In this paper, Bates clearly stated the case for mimicry which was the superficial resemblance of a palatable species (the mimic) to an unpalatable species (the model) as a form of protective colouration evolved through natural selection, positive proof in nature of how species might evolve.

The Biographers

There are few biographers of Bates but I must acknowledge a natural debt to them all. First, there is Edward Clodd whose book *Bates' River Amazons,* was published in 1892. Clodd was an amateur naturalist by inclination, as well as being a friend, neighbour and the executor of Bates's will. He republished the 1863 edition of *The Naturalist on the River Amazons* in its original form, adding an extensive memoir to the book. The importance of the latter is that it is a first hand recollection of Bates by someone who was close to both the man and his family. After Bates's death, Clodd remained in contact with Sarah Bates as her 'friend in need'. The significance of the content of Clodd's book is also that it made the text of the first edition available to a wide readership.

Second is George Woodcock, who in 1969 published the only book-length biography on Bates, titled *Henry Walter Bates, Naturalist of the Amazons.* Woodcock, a Canadian writer, was a professional biographer whose studies included subjects as varied as Gandhi, Aldous Huxley, Kropotkin, Malcolm Lowry, Orwell, Herbert Read, and Oscar Wilde along with studies of anarchism and books about Canada. Over more than 60 years, Woodcock produced an enormous amount of journalism and well over a hundred books. His book on Bates, which is a knowledgeable piece of work, was part of the series *Great Travellers,* of which Woodcock was the general editor.

Third, there is Professor H P Moon, who in 1976 published *Henry Walter Bates FRS, Explorer, Scientist and Darwinian 1825-1892.* Professor Moon, then Emeritus Professor of Zoology at the University of Leicester, re-emphasised the importance of Bates the scientist, highlighting the Leicester link. Though short, it is nevertheless a main source on Bates and lists his achievements chronologically.

Finally, there is Monica Lee, whose book *300 Year Journey, Leicester Naturalist Henry Walter Bates, FRS, and his family, 1665-1985* was privately published in 1985. Lee, a distant relative of Bates, wrote this manuscript on the Bates family, privately printing it because of its proximity in timing to the publication of Professor Moon's book. It is a very important source of information and includes copious details from family sources. Lee is the most important source of information for Bates's early days, providing a comprehensive background to the family and their circumstances, with the all-important family trees clearly described. It is also a source of interesting photographs.

Otherwise, there are brief references to Bates in many other books but usually as a footnote or short paragraph giving him credit for his journey on the Amazons, or for the theory of mimicry.

Social Positioning

The problem for both Bates and Wallace was that they were essentially travelling naturalists and reputation was what they sought. Unlike Darwin, Bates and Wallace had no financial backing and therefore had no real choice other than to be peripatetic collectors. Bates and Wallace both wanted recognition for their scientific discoveries but could only do this if they had acquired the indispensable professional status demanded by the London scientific elite. Eventually both could speak the language of the scientist, but Bates's problem was that after his return from the Amazons, when he was no longer the exotic traveller in the jungle, he could not escape his relatively humble trade origins or hide his lack of university education. Recognition also depended as much on lobbying as on learning and Bates was not good at that either. These factors must have affected his chances of obtaining a scientific appointment and denied him full recognition when he most needed it. Making the local butcher's daughter pregnant and marrying her some time after the birth of a child also spoiled his image

Nevertheless, progress in the application of technical knowledge depends on the use made of the abilities of especially gifted persons like Darwin, Wallace, and Bates, together with those of many who were less able. These abilities include what Bates had in abundance, the passion for inquiry that is the main driving force of scientific advance. However, its scope is limited by the extent to which those who possess it are allowed to express it. Each of these men played a role in the development of science during the nineteenth century and each contributed to knowledge, shifting the position of science in a fundamental way: Darwin with the *Origins*, Wallace with the origins and biogeography and Bates with mimicry. However, there was a natural and rightful hierarchy established in the order: Darwin, Wallace, and Bates. The question is, could Darwin have achieved his greatness without Wallace, or could Wallace have done the same without Bates. In each case the answer is no. Darwin needed Wallace for the impetus to come out of the closet and publish and might even have needed some of Wallace's ideas to achieve this. Wallace needed Bates in his formative years, which were critical for the development of his own ideas on the origins and zoogeography whilst in Malaysia. Moreover, they all used each other as a sounding board for their ideas, with Darwin relentlessly networking them and the system in general. Darwin quickly acknowledged that Bates's extraordinary powers of observation highlighted some of his earliest insights into the complexity of rainforest ecology and because of Bates, Darwin saw how extraordinarily the lives of individual species were linked, confirming how they had evolved to fill precise niches in their surroundings.

The development of scientific ability today would depend on two different criteria. In the first place, anyone with potential scientific ability would be encouraged to use it if a method could be found that would pay for the lengthy education needed for its fulfillment. Secondly, if these particular gifts were directed towards a specific line of inquiry, they would still be unable to apply them fully, unless the government or some other body with funds thought the study in question was worthwhile and deserved support.

Whilst Darwin was a wealthy gentleman, Wallace came from impoverished gentry with little opportunity for betterment and Bates came from a comfortable trade background. Darwin wined and dined Richard Owen, Charles Lyell, Thomas Huxley and Joseph Hooker and counted them as his friends. Bates could claim the acquaintance of these men but not on the same terms as Darwin. Initially he was invited to Darwin's home, Down House, out of curiosity and thereafter only when his scientific opinion was required. However, he conducted the friendliest correspondence with Darwin, often of a personal nature rather than purely scientific.[26]

An incident that tells us something about Bates's relationship with other men of science, and caricatures the self-opinionated Charles Lyell, occurred on a cold November day in the late 1860s. Lyell met Bates walking near the seal pond in Regent's Park and greeted him with "*Mr Wallace I believe*". Bates was annoyed but Lyell apologised and then made the matter worse by saying that he always confused the two of them. After this, they turned and walked together while discussing the success of Bates's book. Sir Charles wanted to know, probably for comparing Bates's literary success with his own, how many copies of *The Naturalist on the River Amazons* had been printed. Bates told him 1250, but Lyell boasted he had done enormously well with his own *Principles of Geology,* which by then was in its tenth edition. The assurance of Bates's more modest success almost certainly satisfied Lyell's vanity. '*Old Murray used to say to me…the more complete you make your book the fewer I shall sell of it*' continued Lyell. The older man's vanity astounded Bates who summed the event up with '*Sir Charles ... has the appearance of a*

---

[26] Darwin Correspondence Project: Extract from Letter 4323: Bates, H. W., to Darwin, C. R., 24th October (1863) 22 Harmood Street, Haverstock Hill, NW.

fidgety man, not well at ease with himself. He is very greedy of fame, and proud of his aristocratic friends ... Darwin says he (Lyell) *likes to hear himself talk.*' [27]

However, we must travel to Leicester in order to discover more about Henry Walter Bates.

[27] *The Naturalist on the River Amazons*, Bates (1863) Clodd, Edward (1892): lxxi - lxxiii.1969).

Chapter 2

Leicester

Leicester had a population of about 40,000 in the 1840s that would rise to 160,000 by the time Bates died in 1892. The town's business was exclusively the making of hosiery in the 1840s, which eventually led to trade stagnation and subsequently left-wing political activities that gave rise to the nickname of '*radical Leicester*'. However, by the end of the Victorian era, Leicester had become a very different place, offering a wider diversity of employment.

In 1840 it was a seedy, overcrowded town with an ill-housed population, but in the late 1850s economic growth led to improved prosperity and much of the underprivileged area was rebuilt. People living close to or below the subsistence level made up a large part of Leicester's population in the 1840s, with about a third of its inhabitants receiving poor relief. The buildings they lived in called '*The Rookeries*' were old, dilapidated and often filthy. Many were lodging houses often harbouring disease and crime. Nevertheless, despite these poor conditions, housing in Leicester was actually better than in many similar Midland towns, as it had no cellar-dwellings, relatively wide streets and numerous private gardens, even in the centre. The town was also surrounded by beautiful and easily accessible countryside.

During the 1840s there was a high mortality rate in Leicester, as in other similar English towns, attributed to poor sanitation and an inadequate water supply. In the 1850s, a virulent form of diarrhoea, evidently caused by blocked sewers, affected both rich and poor alike. The struggle to provide a satisfactory sewage system and clean water for Leicester was not fully resolved until the building of a sewage works that was in use by 1890.[28]

In 1846, Joseph Dare, a minister and missionary of the voluntary Leicester Domestic Mission, set up by the Unitarians to help the poor within their own homes, produced a series of annual reports that aimed to fight the ignorance, filth and fatalism that existed in Leicester. Dare's reports give an insight into the spread of smallpox within the city, with builders, property and workshop owners being criticised for providing insufficient hygiene and ventilation systems.[29]

Leicester was the county market town and it was in the meat market that John Mason first set up his stall as a butcher after he ceased general

[28] Simmons, Professor Jack, East Midlands Oral History Audio Archive, 129, RL100/0001/vols 1-9, 1972.
[29] Ibid.

labouring in the 1850s. Henry Walter Bates was to marry his daughter Sarah. Leicester was also a strait-laced place with early advertisements for trade regarded as morally degrading particularly as they were allegedly boastful.

The appointment of a single Police officer for the town took place in 1836, but Saturday night brawling was a common occurrence in Leicester until the Liberal Government's Licensing Act of 1872 led to calmer streets at the weekend.

Leicester was one of the first towns to have a museum but one of the last to acquire a public library. In 1830, the provision for education was inadequate and entirely dependent on private endowment and benevolence. Religious groups, setting up schools of their own, initially established a good deal of the public education provision in Leicester but, without inspections, few children attended school regularly.[30]

There was a long tradition of the Leicester textile trade in the Bates family and an important link with the cottage industry beginnings of the hosiery trade. Mary Donisthorpe, who was married to Robert Bates, grandfather of Henry Walter Bates, was a direct descendent on her mother's side of William Iliffe who secured his place in the history of Leicestershire by setting up the first stocking frame at Hinckley in 1640. The first record of stocking making in Leicester dates from 1597, when poor children were put to work knitting jersey leg wear, probably using a combed worsted yarn. This remained as handwork for almost a century, as Leicester did not have the stocking frame until 1680. Robert Bates's trade was dyeing and trimming, important intermediate processes in hosiery making. After being knitted, stockings were dyed and then trimmed or shaped using a steam press or former.[31]

Robert Bates was prosperous and appointed his second son, Henry, then 30 years old and a warehouseman, together with William Adams, a friend and hosier, as trustees in his will.

Robert Bates left his entire estate to the trustees to sell by public auction or private contract for the best price they could reasonably get. The sale was to include household goods, furniture, plate, linen, china, stock-in-trade, money, securities, and all personal assets. When everything was sold, the final sum was to be divided equally among his children: Charles, Henry,

---

[30] Simmons, Professor Jack, East Midlands Oral History Archive, 129, RL100/0001/vols 1-9, 1972.
[31] Lee, Monica, *300 Year Journey, Leicester Naturalist, Henry Walter Bates, FRS, and his family 1665-1985.* (Havant: Lantern House, 1985) 6 [After this referred to as M Lee, 1985: no.].

Frances, Ann, Henrietta, Mary, Jane, and Harriett. For Jane and Harriett who were unmarried when he made the will, he set out special provisions:

> "Until the expiration of twelve months after my decease, I will and desire that my daughters, Jane and Harriett shall be permitted to reside in, and make use of, my messuage, tenements, yards, gardens, shops and furniture, ...to carry on my trade or business of a trimmer and dyer, for their own exclusive benefit and advantage, and that they shall have unmolested use of all my implements of trade during the set period of twelve months, but shall, at the expiration thereof, restore and deliver up to my trustees, in as good a state as may be, all the household furniture and implements of trade which may, or shall be, contained in an inventory to be made immediately after my burial."[32]

When Robert Bates died, trustee William Adams was also dead, so it was to Henry that probate was granted in March 1837. By then, Henry had progressed from warehouseman to founder of the stocking formers and shapers business that after the 1856 Companies Act became Henry Bates and Co. Ltd. At the same time, he was involved briefly with the licensed trade, as the landlord of the Rainbow and Dove public house in Northampton Street, Leicester, in all probability to supplement his income as his hosiery business grew.

Not far away from Northampton Street was the newly completed St. George's Church, the first Anglican place of worship built in Leicester since the Reformation. Henry Walter was christened at St Margaret's Church, Leicester, but it was at St George's that Henry Bates's other children were baptised. Despite these Church of England christenings, Henry was a committed Unitarian and the records of the denomination's Great Meeting include a reference to a Mr. Bates of Leicester in the list of members at that time. Whilst there is no initial against Bates to confirm it was Henry, and as Unitarians were commonly involved with the licensed trade, it can be safely assumed it was him. Later, his son Frederick worked in the licensed trade, founding the Leicester Brewing and Malting Co. Ltd.[33]

---

[32] M Lee, 1985. 8.

[33] M Lee, 1985. 9.

Plate 4.
Honest Henry Bates

Leicester's hosiery making was Henry's chief interest, and by the time he died in 1870, he had established a viable company to hand on to his sons.

His will was a simple document stating that all his effects, including the sum of £3,520 in the business carried on by Samuel Bates in the name of Henry Bates, was left to Henry Walter, John Oscar, Frederick and Samuel as tenants in common and in equal shares. The appointment of the executors was subject to Samuel being allowed to buy out the other three brothers should he wish to do so, which in fact he did over time. When John Oscar died five years later in 1875, his estate was valued at £18,000 indicating the growth of the family business over this period. After the Amazon, Henry Walter tried to return to the hosiery trade but decided it was not for him. Samuel bought him out in 1861 for an unrevealed sum (probably £1000) and an allowance of £100 per annum.[34]

[34] M Lee, 1985. 10.

The Early Years

Henry Walter Bates was born on Tuesday, 8th February 1825 at 16 Waterloo Street, Leicester. The house was a solid, pleasant and unpretentious three-storey middle class home of a good standard. Waterloo Street was a short walk from the main London Road, which in 1825 was a busy thoroughfare with a constant flow of horse-drawn traffic near where the railway station was eventually built.

His father, Henry Bates (1794-1870), was widely respected within Leicester's hosiery trade and known as *Honest Henry*. He was a Unitarian in his beliefs and a friend of the nonconformist divine Robert Hall, (referred to as Robert Hale in Professor Moon's book). In 1824, he married Sarah Gill (1803-1860), a spinster of the parish of St Margaret's, Leicester whose father was also in the hosiery trade. She was said to have been both gracious and affectionate but was delicate. The following year their first son, Henry Walter was born. Three more sons, John Oscar (1827), Frederick (1829) and Samuel (1839) completed the family.[35]

Waterloo Street and nearby houses were mostly occupied by manufacturing tradesmen and their families. Many of them were aspiring members of the middle class, forming a friendly, closed, and protective community. In 1973 when the house was finally demolished, progress came in the form of a wide, modern and impersonal main road called Waterloo Way.

Prior to this the Bates's house had been used as a retail shop, a small bakery company, and last of all as the Costa Rica cafe.

Although number 16 was well placed for the town and its amenities, Henry Bates, now a moderately successful businessman, considered he should live nearer to the wealthy centre of the town. He duly built a better home in Queen Street, where by the end of 1841, his ambitions to be seen as prosperous and middle class were fulfilled.

From an early age, Henry Walter and his brother Frederick were fascinated by natural history and in the new house in Queen Street, Henry Bates provided a study-bedroom in which these two boys could use a table that had drawers for storing the insects they collected. The ability to provide enough space for non-essentials clearly indicated a degree of affluence and comfort that Henry was able to provide for his family. It was a sign of both indulgence and foresight, unusual on the part of a man in trade at that time

[35] M Lee, 1985.Genealogical Trees.

who would generally have scorned such pandering to children. Henry Bates, as with many of his generation, anticipated the benefits of education and saw qualities in Henry Walter that convinced him an investment in his education was warranted. He may even have thought Henry Walter was going to do something out of the ordinary with his life and therefore if only one child was to be well educated, it would be him.

Early Schooling

In Leicester the lower strata of the middle class sent their children either to minor boarding schools established near the larger towns, or to one of the many day schools in the town itself. Some of these maintained an existence over a considerable period. The schools varied greatly in quality but most were by no means as bad as the schools criticised in contemporary statements by Charles Dickens. The provincial newspapers of the period were full of advertisements for such local schools. In three issues of the *Leicester Journal* for January 1819, for instance, schools located in and around Leicester are clearly being advertised to appeal to different classes of the community:

> "The Reverend Nicholson advertises his old-established seminary where young gentlemen are expeditiously instructed in every branch of classical, polite and useful literature, as may best suit their future destination, whether the Church, Army, Navy, Commerce, or the more retired scenes of private life, at a cost of 30 to 35 guineas per annum, or as parlour boarders, 50 guineas."

> "The Classical and Commercial Academy at Billesdon, conducted by a dissenting Minister, (Dr. Creaton) announces its advantages in terms more carefully calculated to attract the middle class. The establishment 'is patronised by gentlemen of high respectability... instruction is communicated on an improved system, which has been tried for years with success, which expedites the student's progress, and which embraces every kind of education usually in request. While suitable exertions are made to promote the improvement of the pupils in French, and in classical literature, those parts of learning more necessary to trade and in commercial pursuits, receive a large share of attention. Many young gentlemen have left the seminary highly accomplished in English grammar, a qualification of peculiar

importance in every respectable station in life...The situation is retired, pleasant, and healthy. Terms, 21 guineas per annum."[36]

An analysis of the curriculum showed that the school provided teaching in reading, writing, grammar, arithmetic, geography, history, mathematics, and languages, up to the age of about fourteen. It also taught natural history, drawing, and a little moral and religious teaching.

When the time came to decide upon a school for Henry Walter, his father chose Creaton's Academy at Billesdon, a large village nine miles from Leicester. Creaton was a Dissenter, something that would also have found favour with Henry Bates. It has often been observed that the growth of industry was connected historically with the rise of groups which dissented from the established Church in England. Many explanations have been offered for this close association between industry and Dissent. It could have been that those who sought out new forms of worship would also naturally strike out new paths in secular fields. Equally, it could be that there is an intimate connection between the system of belief peculiar to Nonconformity and the rules of conduct, that lead to success in business. It could also be asserted that the exclusion of Dissenters from the universities, office in government and administration, forced many to seek an outlet for their abilities in industry and trade. There may be something in each of these contentions, but a simpler explanation lies in the fact that the Nonconformists generally constituted the better educated section of the middle classes. This is what Bates senior wanted for Henry Walter.

He would also have been impressed by the school's syllabus, a combination of classics and commerce with an emphasis on English grammar. Sport was part of daily life at the school but Henry Walter much preferred indulging in his passion for entomology, to games. He found the quiet lanes and hedgerows near Billesdon more rewarding hunting grounds than the school playing fields, as they contained insects that he had not seen closer to home.

His brothers did not attend the same school and missed him greatly during term time. Years later in a remarkable sentence speaking volumes about the brothers' relationship, Frederick was to say:

---

[36] Simon, B, Ed. *Education in Leicestershire. A Regional Study*, (Leicester: Leicester University Press, 1968) 117.

> "We used to look forward to the time when holidays would bring him once more amongst us, for even in those days we looked upon him as our dear guide, philosopher and friend, who ever sought to lead us, in his kindly and genial way, to better and higher thoughts and deeds."[37]

Henry Walter was a model student and extended his studies whenever the opportunity arose. He was hungry for learning and had a considerable flair for languages, his natural history giving him a basic understanding of Latin. He never lost his ability with languages and many years later whilst in the Amazons, easily learned German and Portuguese.

Bates left school just before his 14th birthday, as there was generally no provision for schooling beyond that age and it was the custom for tradesmen's sons to be articled. Bates, destined for the family business, was apprenticed to Alderman Gregory, another hosiery manufacturer of his father's acquaintance, who had premises in Halford Street, Leicester. Bates's work consisted primarily of the unskilled tasks of the workplace, including sweeping the floors of the factory. These were hard times because of the long working hours, from seven o'clock in the morning until eight o'clock each night, six days a week. Bates was sometimes accompanied by Frederick, then aged nine, who would assist with sweeping up and collecting string and waste from the floor, which were the perquisite of the apprentice.

Frederick said, when looking back, that it was then that his brother laid the foundations of the diligent, methodological, and hard working naturalist of the Amazons and the efficient administrator of the Royal Geographic Society.[38]

The Unitarian Background

The four sons grew up in the midst of the nonconforming tradition that flourished among the midland textile trades. Strongly principled, expressing shared aims and experimentally communitarian, Unitarianism was in sympathy with science, and committed to civil and religious liberties.

In this environment Bates was enthusiastic about all learning, reading in bed by candlelight, devouring Gibbons's *Decline and Fall of the Roman Empire*, a book he read and reread. He did not allow his education to cease with his

---

[37] M Lee, 1985. 19.
[38] M Lee, 1985. 18.

departure from Creaton's Academy. He enjoyed music, sang with a glee club as a baritone, and learned to play the guitar. While Bates was still an infant, Brougham founded his Society for the Diffusion of Useful Knowledge, which made available a wide choice of encyclopaedias and factual periodicals. Books of all sorts were becoming cheaper and more readily available, and Bates 'soaked himself in printers' ink'. As he grew up exciting events were taking place including Queen Victoria's succession to the throne in 1837.

The Autodidact

Many adult workers seeking knowledge focused their attention during spare time on the Mechanics' Institutes. In the early to mid 1800s, local clubs and societies where working boys and men congregated and where evening learning took place became less financially viable and more dependent on middle-class patronage. As a result, the locally set curriculum of the Leicester Mechanics' Institute reflected the aspirations of the town's nonconformist textile manufacturers. Local clergy would occasionally censure the Leicester Institute for being a place for encouraging republican and levelling principles. However, there were some inspirational teachers there like a Messrs Riley and Hollings, under whose skill and tutelage the Leicester Institute was to develop into a highly successful educational establishment. Bates studied Natural History as well as learning Composition, Drawing, Greek, Latin, and French at the Institute. On the flyleaf of a book of Latin grammar still owned by a member of his family, he wrote 'I am as fond of Latin as women are of satin', which seems quaint for a young boy, even in Victorian times.[39]

On 8th May 1840, an exhibition opened at the New Hall in Wellington Street, Leicester, to raise funds for the Mechanics' Institute. Admission was 6d and the exhibition included '*Many thousands of preserved specimens of Foreign and English Quadrupeds, Birds and Insects*'. Bates was 15 at the time and with his enthusiasm for natural history, it must be assumed that he saw these exhibits. In 1845, an Act of Parliament allowed local councils to levy a halfpenny rate to establish museums for the '*Instruction and Amusement of the Inhabitants*'. Leicester was one of the first towns to take advantage of this, largely on the advice of the Leicester Literary and Philosophical Society, which was founded in 1835. Like the Society itself, the Museum was a means

[39] Bates, Frederick, Obituary Henry Walter Bates, FRS, *Proceedings Royal Geographic Society*, XIV: 245-247.

of healing some of the political and religious divisions that had marked the development of the town since the late eighteenth century. '*No language which I can use can express the immense debt of gratitude we owe to that small band of men who took part in the original formation of the Society...*', said the Mayor, William Biggs, in accepting the Society's gift of the 10,000 objects which made up its own museum in the New Hall in Wellington Street. Many years later, the Leicester Museum would purchase 200 butterflies caught by Bates in the Amazons, for the sum of £15. However, the politics of Victorian Leicester were lively and very often bitter. It was a stronghold of radicalism and Thomas Cooper, the Chartist, kept a shop in Church Gate. In 1842 and 1850 there were serious Chartist riots in the town.[40]

At the Institute, Bates met an interesting group of young men among whom were John, James and Nathaniel Plant and James Harley. Harley was interested in birds; John Plant was eventually to become the curator of Salford museum, his brother James became a noted geologist, and Nathaniel the first curator of the Town Museum in Leicester. Bates grew intellectually mature in the stimulating company of these young companions. As well as entomology, Bates wanted to learn more about new and different subjects. Botany was one of these, and within a brief time he could name, in Latin, every variety of wild flower he found on his Leicestershire rambles.

Bates realised at an early age that to learn gave him the potential to outgrow the confines of the circumstances that would otherwise constrain him. He worked extremely hard at the Institute, gaining first prizes for Greek and Latin, and second awards for French and essays, as well as becoming a skilled draughtsman. At home he continued with his homework until midnight, rose at four in the morning to read more Greek and Latin grammar, or translate some Homer before breakfast, and then he would set off to work. He often woke Frederick at this early hour so that the younger boy could test him on his learning, but Frederick's response to these early morning calls is undocumented.[41]

Entomologizing

Good Friday, a holiday and the start of spring, was the opening day of the entomological collecting season. It was also the end of the foxhunting season, so there was no conflict of interest. On warm days, the collector

---

[40] Dawson, Jan, Curator of Natural History, Leicester County Museums.

[41] Bates, Frederick, Obituary Henry Walter Bates, FRS *Proceedings Royal Geographic Society*, XIV: 245-247.

might be fortunate enough to see the first yellow Brimstone butterflies that have always heralded the arrival of spring.

Henry Walter and Frederick, often accompanied by the Plant boys and James Harley, would set off into the lanes and woods of Leicestershire for the start of a new season's collecting. Bates wrote accounts of these expeditions, sketching and describing all the most important insects they captured. This painstaking habit of noting down everything of interest was the underpinning of his manner of working for the rest of his life. Two journals of his travels on the Amazon still survive in the Natural History Museum in London. Frederick, whose own entomological work was to be of some note later, no doubt also profited from his brother's example.[42]

In the eighteenth century, butterfly collecting was considered an eccentric pastime for both men and women, but in Victorian England this changed and it became both respectable and fashionable, a trend that was to continue well into the Edwardian period. The serious study of butterflies, as an aspect of Natural History, could also lead to the collector having a more important and beneficial place in society or in scientific employment. What had until then been observed as eccentric behaviour was also seen as the preserve of the gentleman amateur. Entomologists, or Lepidopterists as they were known if they only studied butterflies and moths, could come from any stratum of society and were usually the owners of cabinets in which they kept their prizes. Most were avid readers of the fascicules or book parts of entomological literature that were published at sixpence each week or one shilling per copy if they contained coloured plates. In the 1850s butterflies and moths were plentiful, not as scarce as they are today, and they would have been present in large numbers in the hedgerows and meadows. Due to this and their natural beauty, collecting them was a popular craze. The early collectors of the eighteenth century met at the Swan Tavern in London as a semi-professional group calling themselves 'Aurelians', named after the golden colour of some chrysalises. In the nineteenth century, with the advent of the railways generally making travel much easier, local natural history clubs were popular and sprang up in most towns and in many rural areas. More people with an interest in natural history could now travel by rail to collecting grounds. Gatherings took place wherever specimens were known to be in abundance, particularly the much sought after rarities. It would not have been unusual for dozens of collectors to be present and accumulate many hundreds of examples of a species, something that would

[42] M Lee, 1985. 21.

be unknown and indeed frowned upon today. The New Forest, Wicken Fen, and Royston Heath were some of the most popular locations for these gatherings but the North and South Downs were favoured too. The appeal of the craze, for that is what it was, led to lectures with lanternslides being a popular winter pastime, motivating the listeners to get out themselves into the countryside at the start of the collecting season.

An unattributed Georgian document in my possession, written in a fine copperplate hand, records such a gathering:

> "It is a thing to be lamented that some of the Spitalfields weavers occupy their leisure hours in searching for the Adonis butterfly, and others of the more splendid Lepidoptera, instead of spending time in playing skittles or in an alehouse. Or, is there in truth anything more to be wished, than that the cutlers of Sheffield were accustomed thus to employ their Saint Mondays, to recreate themselves after a hard days work, by breathing the pure air of their surrounding hills while in search of this untaxed and undisputed game. There is my friend the weaver; strong desires Reign in his Breast; 'tis beauty he admires: See to the shady grove he wings his way, and feels in hope the rapture of the day. Eager he looks, and soon to clad his eyes, from the sweet bower nature formed arise Bright troops of virgin moths, and fresh born butterflies, c1810." [43]

The pleasure of butterfly collecting is described in *Nabokov's Blues, The Scientific Odyssey of a Literary Genius*:

> "First, the hope of capturing - or the actual capturing - of the first specimen of a species unknown to science: this is the dream at the back of every lepidopterist's mind, whether he be climbing a mountain in New Guinea or crossing a bog in Maine. Secondly, there is a capture of a very rare or very local butterfly - things you have gloated over in books, in obscure scientific reviews, on the splendid plates of famous works, and that you now see on the wing, in their natural surroundings, among plants and minerals that acquire a mysterious magic through the intimate association with the rarities they produce and support, so that a given landscape lives twice: as a delightful wilderness in its own right and as the haunt of a certain

[43] Author's private collection of books and manuscripts.

butterfly or moth. Thirdly, there is the naturalist's interest in disentangling the life histories of little-known insects, in learning about their habits and structure, and in determining their position in the scheme of classification - a scheme which can be sometimes pleasurably exploded in a dazzling display of polemical fireworks when a new discovery upsets the old scheme and confounds its obtuse champions. Fourthly, one should not ignore the element of sport, of luck, of brisk motion and robust achievement, of an ardent and arduous quest ending in the silky triangle of a folded butterfly lying on the palm of one's hand."[44]

The Bates boys and their friends also went on collecting expeditions to Bradgate Park, a few miles outside Leicester, then owned by the elderly Earl of Stamford, who did not strictly conserve it for game but for foxes. This meant there was easy access to places where insects in their natural habitat were plentiful. For these expeditions, Henry Walter and Frederick carried their collecting nets on their backs under their coats. This curious habit occasionally provoked comment that they could be poachers or even thieves hiding stolen goods. From Bradgate they had easy access to Charnwood Forest, another place that yielded a rich harvest for the intrepid entomologists. By now, entomology was becoming more a way of life than just a hobby for Henry Walter. Though still only a young man, he was corresponding with the top naturalists of the day, which clearly indicated that he not only had the knowledge but also confidence in his knowledge and entomological ability. In January 1843, at the age of 18, he made his first contribution to entomological literature with a paper entitled, *Notes on Coleopterous Insects Frequenting Damp Places* which was published in the first issue of *The Zoologist.*[45]

Learning Commerce

Alderman Gregory died in 1843 before Henry Walter's apprenticeship was completed and Gregory's son, who lacked interest and business acumen, was pleased to leave the running of the warehouse to his late father's apprentice. Dissatisfied with this, Bates left Gregory's to spend a few months first at Messrs Bedell's, then at Messrs Wheeler's as a clerk, both establishments being local textile manufacturers. In 1845, a banking friend called Edwin

[44] *Nabokov's Blues*, The Scientific Odyssey of a Literary Genius. Johnson & Coates (London: 1999) 66.

[45] Bates, H W, Notes on Coleopterous Insects frequenting damp places, *The Zoologist*, 1843, vol. 1, 114 -115.

Brown, who also had an interest in entomology, helped him obtain an office job at Allsopp's Brewery, in Burton-upon-Trent.

Here he had new countryside to explore and look for specimens that might differ from those he had found elsewhere. He concentrated on his beetles, finding his evening expeditions into the woods and lanes counteracted the boredom of his days. The result was that he made new discoveries and published a paper in the Zoologist titled *Coleoptera in the Neighbourhood of Burton-on-Trent.*[46]

Through beetles Bates became conscious of the infinite variability of species, the almost endless modifications of structure, shape, colour, and surface markings that distinguished one beetle from another, as well as their innumerable adaptations to different conditions. Such variability demanded explanation and, later on, it was beetles that triggered Bates's thoughts about the origin of species.

By now, he was already considering wider horizons, but had no real idea of where these might be. According to Canadian historian and biographer George Woodcock, Bates was not a particularly fit young man. This may be overstated but he certainly suffered from digestive troubles and used to brew concoctions of Peruvian bark, better known as cinchona or quinine, to combat what might have been a mild form of acne. He tried to improve what he believed to be poor circulation by massage with coarse gloves and at times seems to have veered towards mild hypochondria.[47]

Alfred Russel Wallace[48]

In 1844, Bates met and became very friendly with a 21-year-old teacher, newly arrived in Leicester to work as a tutor in English and drawing at the Collegiate School. He was also interested in natural history, particularly botany, and his name was Alfred Russel Wallace. The bespectacled son of a solicitor and later known as 'Darwin's Moon', he was the co-founder of the theory of the origin of species. Uncertainty surrounds the circumstances of their meeting but one version is that Bates was leaning over the fence of the library in Leicester when Wallace passed by, they talked and discovered they

---

[46] Bates, H W, Notes on local species of Coleoptera in the neighbourhood of Burton on Trent, *The Zoologist*, 1848, vol. V1, 1997-1999.

[47] M Lee, 1985. 22.

[48] Raby, Peter, *Alfred Russel Wallace, A Life*, (London: Chatto & Windus, 2001).

were both keen on natural history, the scientific theories of the time, and the origin of species.

On 8th January 1823, Wallace was born at Usk in Monmouthshire, the eighth of nine children, including four boys, of a middle-class couple of modest means. His father, Thomas Vere Wallace, had Scottish origins and his mother, Mary Anne Greenell, English. Thomas, a lawyer by training, and sworn in as an attorney in 1792, had never been in practice. The family's income derived mainly from inherited property and money, which meant Thomas enjoyed a life of leisure whilst growing up. His mother came from a respectable English family that had lived graciously but modestly in the area of Usk for generations.

After Thomas Vere Wallace's marriage in 1807 his family's fortunes faltered as he tried his hand at a long series of relatively unproductive and financially disastrous ventures, including the publication of a literary magazine. Of the progeny of the marriage, five girls and three boys, four of Wallace's siblings died before reaching the age of twenty-two. Wallace, like Bates, seems to have suffered from frequent illnesses and was considered to have a weak constitution, but his remarkable exploits suggests otherwise.

Wallace went as a boarder to Hertford Grammar School where he was a reasonably talented student and before long was teaching the younger boys their elementary reading, writing, and arithmetic in exchange for a small remittance in his fees. Privileged to attend a grammar school, he also had access to good reading material supplementing that available to him through his father who was then, temporarily, a town librarian.

1835 turned out to be a bad year for Wallace senior who was swindled out of his remaining property and the family fell on hard times. Young Wallace had to leave school the following Christmas at the age of 14, which placed him in a similar position to Bates. He then journeyed to London to be with his older brother John.

In the metropolis he became aware of supporters of the Utopian socialist Robert Owen and was influenced in later life by this experience. In his autobiography *My Life* he recollects that he once heard Owen speak, and thereafter would describe himself as a follower of the charismatic and paternalistic socialist. Owen's down-to-earth and thoroughly convincing social views were the foundation of the cooperative movement, a forerunner of the trades' union confederation. It also inspired the influential Utopian 'communities of equality' that Owen established at Orbiston in Scotland and

New Harmony, Indiana. Wallace retained an interest in Owenism, speaking and publishing on socialist themes throughout his later life.[49]

1837, he had to move once more, this time to Bedfordshire to join his eldest brother William, who managed his own surveying business. This was another critical formative period in Wallace's upbringing, requiring him to learn surveying that would serve him well in his later travels. However, in 1839, he was apprenticed to a watchmaker but later in the same year was back with William who was now based in Hereford.

Because of the time he spent with his brothers, Wallace became a very practical man with skills that would be useful to him for the rest of his life, particularly in drafting and map-making, geometry, trigonometry, building design and construction, mechanics and agricultural chemistry.

Many of these, particularly surveying, took him into the great outdoors. He was fascinated by the countryside and its natural history, especially botany, geology, and astronomy. At this stage, entomology was not of particular interest to him but he had not yet met Bates.

In 1840/41, whilst working in the area of the Herefordshire town of Kington, the newly formed Mechanics' Institute attracted Wallace's attention as an adult learning centre. How similar this is to Bates's experiences. Later in the same year when he had moved to the Welsh town of Neath, he became involved in the building of the new Mechanics' Institute for the town. When it opened he attended lectures and soon his knowledge and general enthusiasm led to him giving lectures there on various technical and natural history subjects. He produced an essay on the nature of Mechanics' Institutes, which were societies open to all with the purpose of disseminating learning, which found its way into a history of Kington, published in 1845.[50]

In 1843 while Wallace was in Neath with his brother William, his father died aged 72. As a result his mother had to find work as a housekeeper and the family fragmented. Later on that year, the lack of available work forced William to make Wallace redundant and he then immersed himself in the activities of the Institute. In 1844, having decided a regular income was necessary, he applied for a position at the Collegiate School in Leicester, where he was hired as a master to teach drafting, surveying, English, and arithmetic. As always, books were of considerable importance to Wallace who now had access to another good library.

---

[49] Wallace, A R,. *My Life, A Record of Events and Opinions.* 2 vols. (London: Chapman & Hall, 1906).

[50] Wallace, A R, An essay, On the Best Method of Conducting the Kington Mechanics' Institute. (1845) In the History of Kington, ed by Richard Parry, 66-70.

This was when he met Bates, already a talented entomologist, whose collecting activities and specimens captured Wallace's imagination. Their friendship grew rapidly but William died suddenly in February 1845, forcing Wallace to abandon the teaching post he had so carefully secured and return to surveying in order to wind up his brother's affairs. Wallace now had to manage a business not much to his liking and, with the help of his brother John deal with tasks such as debt collection that were anathema to him. All this extra work did not, however, curtail his interest in natural history, which he pursued with ever-increasing vigour as he now considered the possibility of making a living collecting plants, butterflies, beetles and birds. His enthusiasm led to his appointment as the curator of the Neath Philosophical and Literary Institute's museum.

In September 1847 his sister Fanny, recently returned from America, took Wallace and his brother John to Paris. There he marvelled at the insect and bird collections in the Muséum National d'Histoire Naturelle (the ancient Jardin des Plantes). On the return journey they visited the insect room at the British Museum, which impressed him even more. The sheer number, variety, and accessibility of the specimens overwhelmed him.[51]

The Search for the Origins of Life

In 1847, Bates visited Wallace in South Wales, another opportunity for the two of them to spend enjoyable day's insect hunting as well as discussing whether they could make a living collecting specimens for profit. Bates was uncertain; however Wallace's passionate and eloquent arguments overcame his hesitancy. Wallace had run out of ideas for making a living in England and Bates was unhappy with the mediocrity of his life in commerce. The decisive factor in Bates's and Wallace's decision to leave for South America was William Henry Edwards's exciting book, *A Voyage up the River Amazon: Including a Residence at Para.* Born in 1822, Edwards was an American recreational naturalist who had written an account of his travels to Brazil in 1846. He used the kind of prose that could sweep headstrong young men off their feet, and this is just what happened to Bates and Wallace. There appeared to be many wonders in store for future travelling naturalists. Edwards portrayed Brazil as the naturalists' paradise teeming with wildlife and beautiful Indian girls. The towns were cornucopias of the exotic, the

[51] The Musée Nationale d'Histoire Naturelle was founded in 1795 and is located in the Jardin des Plantes in Paris. There are four galleries: the Grande Gallerie, Evolution, Mineralogy, Paleontology and Entomology.

climate the healthiest, and the people eager to assist the foreigner in any endeavour.

Following in the tradition of the romantics like Blake, Coleridge, Wordsworth and Shelley, who 50 years before had sought an intense relationship with nature as a means of escaping the horrors of industrialisation, men of courage and vision could now do the same. Those who were fortunate enough to have a choice, looked with eagerness at the prospect of travel as a means of escape. The importance of Bates's and Wallace's journey to South America and their work there was that it was the underpinning of their role as dramatis personae in the progress of evolutionary theory. When the excitement died down and practicalities were considered, each found £100 towards the cost of the expedition. Bates's father was generous and helped both men financially and they decided to leave England for Brazil in April 1848.

Even though his father had supported the idea, Bates's mother, Sarah objected to the proposal as she did not understand his obsession with natural history or how he and his new friend Wallace could possibly make a living from selling butterflies. Her own health was indifferent and she was concerned about her son's too but she sought the advice of their family doctor who assured her that a different lifestyle and a pleasant climate would probably benefit him. Eventually, with the intervention of Henry Bates, she reluctantly gave the undertaking her blessing. Wallace was invited to stay for a weekend with the Bateses, and the two young men said farewell to their families. None of those present could have had any idea of what was now beginning and where it would all end. The choice of the Amazons was, however, a happy one as, except for the journeys undertaken by the German zoologist Von Spix, the botanist Von Martius, and that of the Frenchman, Count de Castelnau, no exploration of such a species-rich and interesting region had been attempted since the travels of Humboldt.[52]

Collecting would now be their work, as there was a market in London for beautiful new butterflies and strange colourful birds. The plan was to ship their specimens to their newly appointed agent, Samuel Stevens, who would sell them on commission and provide expenses. Stevens was an insect collector himself and the brother of John Crace Stevens, the influential London natural history auctioneer of 38 King Street, Covent Garden. Nothing was too much trouble for Stevens; he bought, he sold, and he

[52] Spix and Marteus explored the Amazon from 1818-1820, and the Frenchman, Count de Castelnau from 1844-1847.

spared no pains to dispose of duplicates to the best advantage. He insured consignments unasked, he could always find the means of getting money conveyed to obscure ports and even made advances against specimens he had not yet received. He was one of those exceptional beings who manage without fuss to do everything a little better than seemed possible. Bates and Wallace had good reason to be grateful to him throughout their journeys. He also kept both men informed about important scientific opinion and advances. Most important of all he launched their names into the scientific circles of London.[53]

Scientific enquiry and approval provided deeper motives than just earning money at this early stage, especially for Wallace.[54]

In 1848 two more European explorers set out to visit Humboldt's 'New Continent'. Since Humboldt, this was a world reborn by the collapse of Portuguese dominance in Brazil, emerging as a buoyant economy with particularly close ties to Europe, as the location of rich natural resources, super-abundant nature, and potential settlement. Following independence in 1822, Brazilians had also fostered an intensifying political awareness that in 1835 erupted into revolt, rapidly setting fires raging throughout the countryside as a chaotic and fluctuating alliance of Indians and slaves plunged the huge province of Pará into rebellion (the "Cabanagem"). Bates, more than many contemporary travellers, acknowledged the continuing shock of this rebellion, referring to it from time to time in his writings.

However, it was to be Amazonian butterflies and beetles that would turn Bates into a leading entomologist. Seeing nature's apparent disarray he set about putting everything in order, using his new understanding of taxonomy, which had evolved in Europe in the seventeenth and eighteenth centuries culminating in Linnaeus's *Systema naturae* in 1753. Bates frequently worked in places where no European scientist had previously set foot, assembling and cataloguing a vast natural history collection dominated by insect and bird specimens, many new to science.

The Radical Doctrine of Evolution in the Natural Sciences

Later Darwin, Bates, Wallace, Hooker, and Huxley were to contribute to establishing the radical doctrine of evolution in the natural sciences. There is

[53] Woodcock, George, *Henry Walter Bates: Naturalist of the Amazons*. (London: Faber & Faber, 1969) 31.
[54] Goodyer and Ackery, Bates, and the Beauty of Butterflies, *The Linnean*, Vol 18, Number 1, 2002, 22.

much to be learned about the workings of British science at this decisive moment from tracing the letters and specimens passing between these and other scholars as they assembled the fundamentals of a credible theory.

Darwin knew of no specific examples from nature to support natural selection but argued his case by analogy. Having spent many congenial hours with pigeon-fanciers and country squires obsessed with the breeding of horses, cattle and dogs, he theorised that nature acts similarly to the breeder who, by selecting certain traits, produces optimum breeds, or a wide range of very different breeds. He even took up pigeon breeding himself and made the rounds of pigeon and poultry shows to discover how natural selection became the conscious agency tirelessly toiling over the perfection of creatures:

> "Natural selection is daily and hourly scrutinizing, throughout the world, the slightest variations; rejecting those that are bad, preserving and adding up all that are good; silently and insensibly working ... at the improvement of each organic being in relation to its organic and inorganic life. Natural Selection would be even more powerful than artificial selection, he believed, for Man can act only on external and visible characteristics: nature cares nothing for appearances, except so far as they may be useful to any being. She can act on every internal organ, on every shade of constitutional difference, on the whole machinery of life."[55]

The only available illustration from nature was the phenomenon of insect mimicry reported by Bates who noted that some insects disguised themselves as natural objects, such as dry leaves or clods of dirt. In one form of crypsis, now known as 'Batesian mimicry', an edible insect has an appearance that copies the bright warning colouration of a poisonous or unpalatable insect and thereby fools its predators. Darwin thought selection was the most plausible explanation for this phenomenon, a theory that would be debated fiercely decades after his death.

In fact, Darwin identified two distinct forms of selection in *On The Origin of Species*. The term 'natural selection' is reserved for any trait that enhances survival, such as superior adaptation to climate, increased resistance to disease, or an ability to evade predators. However, Darwin also saw that an

[55] Darwin, Charles, *Variation of Animals and Plants*, Vol. II, (London: John Murray, 1905) 3.

individual that did not possess superior survival equipment might triumph simply by being better at reproduction. 'Sexual selection' was epitomised for Darwin in the flamboyant plumage of certain birds, the antlers of stags, and other attributes that could influence mating and reproductive success.

'How Stupid of Me Not to Have Thought of That!'

When the great comparative anatomist Thomas Henry Huxley, who became Darwin's best advocate, heard Darwin's ideas, he exclaimed '*How stupid of me not to have thought of that!*' Here was a grand unifying principle linking all life forms from protozoa to butterflies, and its essence was so simple and transparent that even a child could grasp it. There were only two basic tenets: that there is heritable variation, and differences in survival and reproduction among the variants. Some organisms survive and leave more offspring than competing organisms, and the winners' genes are passed on preferentially to future generations.[56]

Darwin conceived his theory in 1839 but did not publish for another twenty years, becoming preoccupied for eight years writing a monograph on barnacles. He had intended to produce a huge volume about his evolutionary theory, and doubtless would have gone on writing and rewriting for many more years if Wallace had not independently developed essentially the same theory based on his observations in the Malay Archipelago, and written to Darwin seeking his opinion. Darwin could afford to delay no longer, and on 1st July 1858, papers on evolution by Darwin and Wallace were read before a meeting of the Linnean Society in London.[57] (Commonly spelled either as Linnean or Linnaean).

On the Origin of Species by Means of Natural Selection, or the Preservation of Favoured Races in the Struggle for Life, essentially only an abstract of Darwin's unfinished book, was published on 24th November 1859, and booksellers snapped up the 1,250 copies of the first edition on the first day. John Murray, its publisher, was a sceptic about evolution but the publication quickly came to the forefront of Victorian consciousness. The disagreements it stirred crystallised in the famous debate in 1860 between Samuel

[56] Darwin F, *Life and Letters of Charles Darwin*, ed. Francis Darwin, Vol. II, p 197, (London: John Murray) 1887. Huxley's famous response to the idea of natural selection was 'How extremely stupid not to have thought of that!' However, the correctness of natural selection as the main mechanism for evolution was to lie permanently in Huxley's mental pending tray. He never conclusively made up his mind about it, though he did admit it was a hypothesis that was a good working basis.

[57] Loewenberg, Bert James, Darwin, Wallace, and the theory of Natural Selection, including the Linnean Society Papers. (Cambridge, Massachusetts: Arlington Books) 1959.

Wilberforce (Soapy Sam), the Bishop of Oxford, who represented the creationists' viewpoint, and Thomas Henry Huxley, who appointed himself Darwin's 'bulldog' and represented the evolutionists' point of view. Huxley, an atheist, had launched his career as a naturalist on board *The Rattlesnake*. He became a brilliant comparative anatomist and quickly embraced Darwin's ideas. Huxley coined the term '*Darwinism*' and at the debate held at Oxford's University Museum it was reported that a few women fainted at some of the heretical ideas expressed in the room. Wilberforce is supposed to have asked Huxley if it was through his grandmother or his grandfather that he claimed descent from a monkey. Huxley reportedly responded in the following vein:

> "...If the question put to me is would I like to have a miserable ape for a grandfather, or a man highly endowed by nature and possessed of great means and influence, and yet who employs these faculties and that influence for the mere purpose of introducing ridicule into a grave scientific discussion - I unhesitatingly affirm my preference for the ape."[58]

After this, undergraduates reportedly leapt onto their seats to cheer, and poor Captain Fitzroy, who was in the audience to support the theological literalists, stomped around the room holding a Bible over his head.

Social Mobility

This was the atmosphere of change that challenged the establishment when Bates eventually emerged from the Amazons. He clearly hoped for recognition of his achievements in science as this was an accepted path of social mobility at this time of unprecedented upheaval during the industrialisation of British society. He had set out in his early 20s with little formal education, few connections, and a paltry sum of money, making his journey to the Amazon an incredibly audacious venture. At this time the established scientific hierarchies sanctioned a clear and subordinate role for self-educated enthusiasts like Bates, mere amateurs lacking the cultural capital to penetrate their elite institutions. There was little encouragement either for theory making. Collectors like Bates were seen simply as a means of supplying armchair savants in the establishment with exotic specimens to

---

[58] Slotten, Ross A, *The Heretic in Darwin's Court*, (New York: Columbia University Press, 2004) 175-178.

fill their natural history cabinets. Bates knew however that a private collection was a recognised form of capital in appropriate circles, had a significant exchange value and indicated status that should have catapulted its owner into the ranks of the accepted.

Chapter 3

In the Footsteps of Henry Walter Bates

Arriving in South America accompanied by my son David, we approached Belém as they cleared away the refreshments in our Varig aircraft, 5,000 feet above the river. It was misty below and at this hour we felt we had come to a strange and dangerous land. The country far below was in a gentle haze, and it seemed silent and empty, yet profoundly mysterious. I had been reading Bates's book and it was odd to sit thinking about his words with the forest beneath us, while sealed in a cocoon of luxury and ease, and to know that somewhere down below, in the early morning, lay the mission of our journey.

Then suddenly we were in Belém and Belém was abruptly in the sun. This was to be our first, last and most enduring impression of this shining continent: the sun. From it, or perhaps from the union of sun, river and sea, springs the whole Brazilian temperament. This country is peopled on the rim round a wilderness of forests and rivers, the rim fringed by miles of coast lapped here, pounded there, by the southern Atlantic Ocean. As for Belém, throughout the year it enjoys an average of nearly seven daily hours of sunshine, and this to Englishmen, equates it with paradise. Nearly everyone is richly coloured, well nourished and informally clad and on the famous beaches around the rim, scarcely clad at all. Hair is gloriously black, children wonderfully joyful and healthy. Here on the 28th May 1848, when the Barque *Mischief* anchored off Salinas nearby, Bates and Wallace viewed the distant horizon from the deck through the Captain's telescope and saw the sparkling sandy brown river for the first time.

Since then modernity has scarred the landscape. Poles and pylons, cables and TV masts, roads and hoardings, shops and signs, straggle up and down this irregular, indented, wooded coastline for a few miles in both directions; but the river remains, the forests, the foam-creamed beaches and the vultures, the heat and the humidity. Ferryboats, tankers, tugs, and liners compete in the harbour, which deals with many thousand boats a year. In the old quarter of the city survive a few of the distinguished buildings put up by Portuguese colonists, and those erected by the rubber Barons, mostly lost now amid mould encrusted edifices of damp modern concrete towers. However, for the most part Belém is an amorphous, undistinguished sprawl of corrugated iron, brick, and concrete interspersed by rivers of water and streams of traffic and everywhere glorious palm and mango trees. Streets are thronged, shops packed, parking spaces full and everyone seems to be journeying with a purpose.

Bates first impressions are interesting for to begin with he was clearly disappointed; he asked where the dangers and horrors of the tropics were when he wrote home to his friend Edwin Brown. 58 Wallace was to say in the same manner the weather was not so hot, the people were not so peculiar, the vegetation was not so surprising, as the glowing picture I had been conjuring up in my mind during the tedium of the sea voyage.[59]

Bates goes on to say in his narrative:

> "The number and beauty of the birds and insects did not at first equal our expectations. The majority of the birds we saw were small and obscurely coloured; they were indeed similar, in general appearance, to such as are met with in country places in England...Occasionally a flock of small parakeets, green, with a patch of yellow on the forehead, would come at early morning to the trees near the Estrada. They would feed quietly, sometimes chattering in subdued tones, but setting up a harsh scream, and flying off, on being disturbed."[60]

Three days after our arrival we travelled by boat to the Island of Marajó . I was escaping the city by-going to see a place that would bring me closer to the Amazon and its forests and nearer to the experiences of Bates and Wallace in the early days of their travels. When we set off the herringbone sky was just tinged with pink. The harbour was full of long wooden boats piled high with bananas and vegetables and as the sky became a deeper pink, black vultures arrived to scavenge in the glistening mud for rotting vegetable matter. It was a scene of picturesque squalor. We sailed for about three hours out into the estuary which is more like a sea than a river, eventually disembarking, and then travelling overland to Rio Camará, a tributary of the Amazon in the delta. From there we were taken in a motorised canoe to a Fazenda called Sra do Carmo, on the banks of the Rio Camará on part of the Island that had been cleared before Bates time for Buffalo ranching and remains today much as he would have seen it. Extracts from my travel diary describe my first steps into Amazonia:

---

[59] Bates, Henry Walter, *The Naturalist on the River Amazons. A Record of Adventures, Habits of Animals, Sketches of Brazilian and Indian Life, and Aspects of Nature under the Equator, during Eleven Years of Travel.* Chapter 1.

[60] Ibid.

"The following morning we went on the river before breakfast, making those first tentative explorations into the unknown just as Bates and Wallace had done in 1848.You draw back the curtains on a rectangle of forest at dawn and realise that you are adrift from everything that gave you identity. You are in a place that is sparsely populated and densely wooded, with a rich flora and fauna. Tall snowy egrets stand sentinel along the water's edge, cormorants keep guard on overhanging branches, and kingfishers dart for fish. Parrots and howler monkeys are in the trees. Thousands of miles from anyone who knows you, you have the illusion that your past is scarcely yours at all. Dangerously, you feel invulnerable fearing only your failure to reach where you are going."

"The forest was suffused with a soft gold light as we made slow progress through the water. The atmosphere felt close and sleepy, the low hum of our outboard engine a monotonous lullaby in the early morning warmth. The channel was getting narrower as we brushed under low-hanging branches and the aerial roots of fig trees, which looked like long dreadlocks tinged with henna. As we sailed quietly on, a noisy troop of squirrel monkeys leapt along the branches overhead. Animals that can fly or climb above the watermark have no need to move out when the waters rise, so the canopy is full of monkeys. They were not the only creatures watching us carefully. The water was infested with Caimans, (alligator-like creatures); small ones mostly but there were the occasional larger brutes lying dead still in the shadowy vegetation overhanging the bank. One slipped into the water as we passed, a monster some four metres long. Black caimans are fearsome predators, active mostly at night and endowed with acute sight and hearing. They feed on fish such as piranha and catfish, and aquatic animals such as the capybara and the giant river otter."

Bates was startled by his first encounter with the Caiman:

"On a muddy islet lay our first large Caiman, a scowling log of cunning malice watching us as we glided along in our canoe. When we got within twenty yards of the reptile it raised its head and dragged itself into the water with a sluggish movement: a muddy splash. It seemed at odds with what appeared such a docile scene to find these great monsters there."

"Not far from the Caiman several jacanas, of the Rail family, heedless of our approach or possible danger, were gambolling on the water-plants. These lively, bright brown birds, yellow under their wings, are the embodiment of grace and high spirits. Nature has endowed them with abnormally long, oddly shaped claws, with which they can support themselves on the surface leaves of aquatic plants. The jacanas ran gaily from leaf to leaf, paying little attention to our presence. Then a brilliantly coloured kingfisher flew over us and alighted on a branch not far from the jacanas. Now another bird began to announce its presence near at hand: tap, tap, tap, beating like a hammer on a tree. It was a yellow woodpecker well hidden from our view so, curious, we floated in its direction. Clinging to the trunk, it pecked industriously at the bark; then flew off into the forest."

"On one untouched stretch of silver sand a white-necked heron was hunched over, subduing a catfish, its long grey feathers ruffled like a scruffy old coat. A few feet behind it, a large black Caiman basked in the sun. It eyed us as the current brought our canoe close alongside. It was a formidable looking animal, about 2.5 metres long, its tail jagged like the blade of a chainsaw. It opened its mouth to expose a row of lethal teeth, and then suddenly it moved forward and slipped into the deep brown water, disappearing under the surface somewhere in front of us."

We stepped ashore just as Bates had done. A cicada sounded like a can of nails in my eardrum, which shaken slowly, then faster and faster stopped unexpectedly leaving the ear tingling as if shattered by a loud bang. Huge leaves fell to the ground swirling about like a helicopter out of control but landing with a faint soft sound. The forest seemed peculiarly dark and lonely, the sun spots rare. Little ghostly brown butterflies flitted over the leaf litter settling seemingly invisible on the ground. Like Bates, we searched for them in vain.

The sight of butterflies enthralled Bates and Wallace as they first cruised on the river system of the Amazons. Wallace never lost the excitement of catching a rare butterfly and describes his feelings when writing from the Malay Archipelago many years later as he netted the bird wing butterfly, *Ornithoptera priamus*:

> "The beauty and brilliancy of this insect are indescribable, and none but a naturalist can understand the intense excitement I experienced when at length I captured it. On taking it out of my net and opening the glorious wings, my heart began to beat violently, the blood rushed to my head, and I felt much more like fainting than I have done when in apprehension of immediate death. I had a headache for the rest of the day, so great was the excitement produced by what will appear to most people a very inadequate cause."[61]

Every now and then butterflies fluttered straight past me, pausing briefly or not at all. Some of their colours were startling and varied: bright yellow, blue and green, crimson and purple. Some were brown and others transparent. Some flew very fast, others floated effortlessly without it seemed any motion of the wings, I never saw any settle for very long, but occasionally one would and then lay an egg on a leaf before continuing its meanderings. When they did settle they uncoiled their long tongues and sipped briefly from hidden juices, and sometimes from bird-droppings. They appeared to be extraordinarily light; indeed they were nearly all wing, all-possible weight pared off so their lift-drag ratio must be almost perfect. A huge blue butterfly, *Morpho achilles* flew about our canoe and in and out of the forest. Various butterflies floated past in ones and twos; monkeys and strange birds emerged and disappeared overhead.

Wilfred E Norris explains the intricacies of catching these magnificent butterflies in Chamber's Journal, (July 1855). The method used was to attach a damaged butterfly as a decoy on the end of a long stick held in his left hand and to wave this about in the air as if the insect was in flight. This would attract any passing *Morpho* that would then be caught in the net held in his right hand. He continued:

> "The air was filled with the beautiful iridescent sheen of the *Morpho menelaus, Morpho adonis*, and other rare butterflies, darting here, there, and everywhere. They soon attacked my dummy. At first, I damaged quite a few by not being quick enough at the kill (capture), but soon I got the knack, and my spoils began to grow…I sat on a fallen tree-trunk at the edge of the stream. The exertion and excitement of the

[61] Wallace, A R, 1869. *The Malay Archipelago, The land of the Orang-Utan and the Bird of Paradise, A Narrative of Travel with Studies of Man and Nature.*

morning had tired me. Henri (his partner) was busy preparing further dummy flies for catching the *Morpho rhetenor*, a similar species to the *menelaus*, but a more intense blue. The capture of these...offers more difficulty, first (because of) their more rapid flight, secondly because they fly at a height of about 25 feet. The larger the area that has been cleared, the lower they fly, but never below 15 feet. The hunter therefore builds a scaffold or platform, called a mirador, and on it he stands and operates in the same way as for other Morpho, the net in one hand, the bait in the other."

"Having, however, such a limited space to work from, he frequently in his eagerness to catch a fly leans out too far, with disastrous results...These high fliers start appearing about ten-thirty in the morning. Besides *rhetenor*, already mentioned, *cypris, hecuba, perseus*, and *metellus* are also captured from a mirador. After one o'clock none of these will be seen again that day."

Once again on the river David and I drifted along the waterline disturbing butterflies, which whirled in the slight breeze like handfuls of confetti. They tested the mud at several points and re-gathered, one after the other, on some particular patch. They jostled, flapping and pushing each other's wings at first, and then they became still, their wings half closed, mostly pearl-white, apricot or lemon tints, and a few vivid colours against the black. Packed so densely on the mud their tongues criss-crossed each other as they tried to feed. They opened their wings slowly when there was room, and then snapped them shut again. Sometimes a butterfly crossed from one little cluster to another; there was no other disturbance, and none ever settled long on the empty expanses of mud between these clusters, which were equally damp. They could be picked up by hand by their wings and when released would return straight to the mud on which they fed so intently. If the whole cluster was driven off, they swirled around before settling again in the same place, probably drawing sustenance from some animal's urine deposit.

Then a gaudy *Heliconius* butterfly with yellow spots and red bands on its black wings fluttered by and began diligently to deposit eggs on the leaves of a *passiflora* bush near the riverbank. Bates had watched these same species in his day in a similar place. We tried to film it without much success. It performed its task with much care for its future progeny. In a fortnight, small caterpillars would crawl out of the eggs, basking in idyllic bliss as they ate voraciously in the warmth of the tropical sun. More clusters of yellow butterflies were all along the water's edge, dipping their long tongues into the moist sand exposed by the falling river. Among them were larger ones of a rich reddish orange, and a single day-flying moth, *Urania Leila*, a wonderful insect of velvet black with emerald green ripple marks on its wings, and long silver tails. We reluctantly climbed back into our canoe.

In the distance the forests gave out a low resonant noise in the early morning and again at dusk, like a gale in the mountains, or heavy seas at a distant beach. This was the awesome sound of howler monkeys chanting, probably a territorial tactic to keep part of the forest for their own. Bates and Wallace were as surprised as we were at this almost indescribable and somewhat frightening noise. The monkeys were silent after the sun had risen and the blue green faded from the grass. There were supposed to be peccaries (pig like mammals) about, though it was uncertain where they would be hiding this morning. We closed in along narrow channels but without success, and looked casually in others. Small Islands, pieces of the riverbank that had broken off, surrounded us; each of these islands concealed a pocket of the land, floating where the streams ran.

Amazonia is a vast forest and a giant river system and to see the former today you need to be above the forest; to see the latter you must be on the river. A sense of otherworldliness grows the longer you spend in the Amazon. You feel not so much a visitor as an infiltrator. It is one of the oddest places in this strange land: an unconventional collection of fauna and flora. However, the whole place is a study in fragility, from the shifting sands that constantly change the shape of the riverbank, to the delicate forests that rest inshore. You may have gazed in wonder out of the aircraft window at the dense never-ending forests and silver ribbon rivers, lustrous in the sun, but at ground level it is even more magical. Meander along the frontier between the land and the water and the sky provides your entertainment. The clouds test all kinds of shapes, as if trying to create an extra dimension to make up for the forlorn uniformity of the forests. Unseen birds supply the soundtrack. We sailed 1100 km from Belém near the Atlantic Ocean to Manaus in central Amazonia to forge a closer affinity with the river and Bates. On a slow steamer travelling at no more than eight knots, the trip lasted six days.

Going upriver is better for observing the jungle at the water's edge because the boats hug the shore to avoid the current. Going down the middle of the river is better for speed and desirable if one's tolerance for high-density living, the unchanging monotony of the forest scene, and the repetitious character of the food, are the criteria. Travelling by boat is still the only way to experience the Amazon and the only reliable method of travel over the long distances separating the main towns. Although there are a growing number of roads through the Amazon, rivers remain by far the most practical form of transport for most people. Local people, and the few intrepid tourists, use the picturesque, cheap, and in many ways practical gaiolas that ply the waters of the Amazon River system.

Although accommodation on board these riverboats is extremely basic, the open-sided decks festooned with hammocks are a logical solution to their slow pitching pace and the sultry, humid climate. The movement of the boat keeps the insects away, and there is plenty of room for passengers. In the event of an accident, it is also easier to abandon ship without the walls and doors of a conventional cabin - a consideration that is all too important on the unpredictable waters of Amazonia.

There is no better way to appreciate the character of the rural people of the Amazon than by travelling with them aboard a boat called a *gaiola.* The name means birdcage, and comes from the cage-like sides of the open decks, but the name is also often appropriate to the long, intricate songs about ordinary people, as well as heroes, which the passengers sing to while away the hours.

On any gaiola, it is advisable to take your own food supplements, and it is essential to carry plenty of bottled water. The food served on board is usually basic, often badly cooked, and occasionally inedible. It usually consists of beans, rice, farinha (manioc flour) and bits of fish or chicken. Toilet facilities are limited and basic, and by the end of the journey can be extremely unpleasant

Despite the hardships, this is the best way to meet the people of the Amazon under conditions that almost force some exchange of culture and experience, regardless of language barriers. In addition the traveller experiences the closeness of the river and the forest at their most magical, especially at night, and is obliged to come to terms with the pace of life in Amazonia, which is somewhere between very slow and stationary.

Floating on the Amazon on a riverboat may sound idyllic but reality is somewhat different. To begin with imagine being without access to a cabin and using a hammock and able to touch a dozen other hammocks without moving yourself. The artificial privacy typically offered by a hammock is obliterated by all the people and baggage jammed aboard the boat. Hammocks are strung no more than 18 inches apart and interleaved both N to S and E to W. Jostling for airspace is then a way of life as passengers come and go, one stop to the next. Notes from my travel diary tell the story of my journey up the river and give a flavour of how the same journey must have been for Bates.

15th October 2003[62]

We went to the docks after lunch to await departure of the boat between 3 and 4 pm. Many multi-deck riverboats are tied up beside us, each with a high-powered stereo system on its top deck blasting out samba music. Our craft, the MV Santarém is relatively new and said to be one of the best on the river. It has three open decks with the middle one allocated for hammocks and the bottom for cargo. The top is reserved for the cabins with an open space for recreation. The bar was down below with the cargo. The back half of the top deck is roofless so people can sit out or dance to the pounding music from portable radios brought on board by the passengers. This is Brazil!

There are ten cabins forward on the top deck. Occupying one of these gives access to your own private lavatory, something to be envied once you have experienced the public lavatories on the lower decks. We secure our gear and watch the steady influx of passengers and cargo. More people mean more hammocks, occasionally leading to territorial disputes as the density of bodies in crisscrossed layers of hammocks increases. Food and beverage vendors are everywhere, hawking last-minute supplies.

4 pm comes and goes but the boat remains moored. When it leaves the boat is amazingly slow. If you stand on the bow, you only feel a moderate breeze. We eat our first chaotic meal where seating is allocated like musical chairs. More testing is the fact that there are only two bathrooms for each gender on deck. This becomes a challenge the next morning for those occupying hammocks when the Brazilian men decide to shower despite a queue of 50 others urgently needing to use the toilets. We have a cabin affording us both some security and comfort but it is stuffy and small. We have a small window through which plenty of fresh air could flow but we are advised it should be kept shut for security.

The shore was very close, giving us a detailed look at the vegetation along with an occasional hut or house. Canoes, typically containing women and children, paddled out to solicit gifts. It must be a Brazilian habit to put something in a plastic bag so it will float and toss it overboard to these

[62] Author's Amazon diary, *Sixty days on the river*, 2003-2005.

isolated people. Looking up the channel, I could see several canoes waiting their turn to paddle enthusiastically alongside as our boat passed.

Last night the boat passed through the Narrows, it was extremely dark and I wondered how the man at the helm could steer. The river has no navigational markers that I can see and few lights on shore. Of course, the boat carries radar. When I stand on the bow Titanic-style, I can see nothing in front of us but darkness and I can feel insects blindly crashing into me. The boat has a searchlight the helmsman flips on periodically to scan the water ahead for debris. In the Narrows, I notice he flashes the shore from time to time as well.

16th October 2003

When it rains, as it does frequently the crew drop blue tarpaulins over the railing to stop the downpour soaking those occupying the hammocks. Not helping matters was the fact that the overhead lights on the boat stayed on all night and that the drunks in the bar conducted an enthusiastic demonstration of stomp dancing or something similar for the duration. Insects of all shapes and sizes swarm to the bright lights and fall helpless on their backs on the deck.

Storms come and go, rainbows appear and fade away, and cats, dogs, and children play underfoot. For the majority there is a complete lack of personal space, with overhead pipes strung with clothes and plastic food bags. Could not help noticing a contingent of very young women in halter-tops and extremely tight shorts loitering on deck. People read, sleep, play cards, and talk in small groups, both in and out of their hammocks, a panorama of puffy white clouds, muddy brown water, and the distant green shores slide past. At lunch, we have our first hearty meal after the hammocks and baggage are cleared from the deck. Long tables are lowered and people quickly queue for a seat on the benches beside each table. It is strictly first-come, first-served but the shifts go quickly. Chicken chunks, beans, rice, and spaghetti arrive in large bowls and are passed around. Rather heavy fare so I am glad we have some supplementary snacks like a jar of Palm hearts.

As cabin passengers, we have the luxury of a small dining room but the food is the same and the serving no different. The boat announces its arrival at a river town with a long blast of its horn. Heavy ropes are tossed ashore and the boat swings in to the dock. I watch food and beverage vendors swarm aboard as passengers disembark and cargo is loaded or unloaded.

Hardworking black stevedores, naked to the waist, carry everything from beer, rice, and salt to refrigerators and truck axles. We stop at another small town to drop off a woman who leaves our view in the back of a pickup truck. The boat remains tied up for the night as nonstop Brazilian pop music blasts from the upper deck where people drink and dance under the night sky. Unable to sleep we migrated there too and joined them as lightning flashed on the horizon.

17th October 2003

The boat stopped at a place called Juruti. It has two main streets that intersect in a T at the dock, which is a concrete slab 50 meters long. Add a dozen or so simple shops, a smattering of muddy cars and trucks, and a small church at the far end and your picture of a jungle outpost is complete. Last night we saw an enormous fish being carried aboard. As the day wears on, the idleness of boat travel is enhanced by the monotony of seemingly going nowhere. Other single and double-decker boats come and go but ours sits there, hogging the dock. People are uncertain about how far we can wander from the boat as exploring the town carries the risk of being left behind.

Expecting fish, certainly one of Amazonia's favourite foods, people anticipate meal times for lack of anything better to do. Hovering over an occupied bench to grab a seat as soon as someone finishes eating becomes common practice. Even though you must then sit amongst the carnage until the table is cleared and wait for new plates and utensils, the large bowls of beans, rice, and spaghetti along with a heavy platter of grilled fish is worth it. There is usually plenty for everybody after the initial anxiety subsides.

A heavy rainstorm arrives in the afternoon but we lower the tarpaulins on the side of the deck in plenty of time. The flat terrain and open sky makes it impossible for the weather to surprise us. As the rain hammers down, there is talk of stripping down to take a shower.

No one wants to use the showers on the boat since we know that all water has a common source. One man's toilet is another man's shower, the light brown colour of the water and its smell tell the tale. Later in the evening, just as the boat is getting underway, another 'wall of water' rainstorm engulfs the town. It trapped a contingent of bored male passengers who had gone ashore to drink in the local bars. When the horn blasted, everyone sprinted for the dock, getting completely drenched in the process.

18th October 2003

Arrived at Santarém at dawn and the usual chaotic transfer of people and cargo began. Defending hammock spaces seems to be a priority when docked. The boat is now considerably more crowded with hammocks in every nook and cranny, especially the eating area (which must be taken down or rolled up when a meal is served). Breakfast is a buttered roll and fresh fruit with sweet, milky coffee.

By 2 pm everyone now wants (and needs) a shower but is afraid that the harbour and its water is the source of the stink, not surprising when you realize that every boat there dumps its waste overboard. Not even the showers reserved for the cabin passengers are exempt from this risk. One passenger and his new girlfriend (one of the young women with painted-on shorts) were nearly left behind. Out on the river the breeze was stiff, removing all trace of the dirty feet smell so the showers became quite busy. Millions of yellow butterflies fly over the boat from north to south in a continuous stream all day and well into the night and continue to do so for remainder of our journey. (A phenomenon known as *panapaná*).

19th October 2003

The day broke grey and misty. The thin green line comprising the shore is almost out of sight now that the sluggish brown mass of the Amazon has widened. Clumps of vegetation and occasional trees float downstream past us. From time to time the boat passes large islands with dense vegetation growing down to the edge. Houses of the river people appear on the banks, proof that Amazonia provides enough for the resourceful. In the far distance there are hills with flat tops at a place called Almeirim. The day continues wet and grey. The girls with the painted on shorts continue to flirt with the boys. What I would have given for a Portuguese phrasebook. It was a good day to do absolutely nothing; David would have liked to sleep in a hammock, which by day feels quite comfortable as it swings with the rhythm of the boat. It is always much easier to fall asleep in a hammock when you do not have to.

20th October 2003

Little sleep last night and after the breakfast I watched more children from the nearby shores paddle out to beg for treats or try to hitch a ride by looping a rope around a bit of railing or a tire being used as a fender. Rather dangerous but then so is living in the rainy jungle. Eventually the boat reached Manaus, one of the rainiest cities in the world where there is no dry season. The big soft bed and cool air conditioning at our Hotel more than compensated for the gently swaying boat and sounded not unlike a faintly throbbing diesel engine.

Some days later after a storm, we had travelled nearly thirty miles up the Rio Puraquequara (A Tupi word meaning "the place of the electric fish"), a river running parallel to the Rio Negro and a tributary of the Amazon and it was here that we were introduced to the Amazon fauna. A giant tarantula parked itself on the pathway near the thatched cabin we were occupying, perhaps attracted by our lamp, or disliking the damp of the drenched forest, or both. It was about the size of a small soup plate, and was covered with hair, which appeared to be standing straight up, as were my own. These hairs, I learned later, can be dislodged and can cause great irritation and discomfort to anyone they touch. There are a number of species of Amazonian tarantulas, but none of them, unless actually severely provoked is disposed to bite humans. This was something else we quickly learned, that things were significantly less dangerous than we had originally thought. This incident would hardly be worth noting, except that it revealed a confusion shared by most people, myself included, about the nature of this rainforest wilderness. To many the very name Amazon suggests a Green Hell: forty-foot anacondas, piranha fish, which will strip the flesh from the bones of man, dreadful terminal fevers, ulcerous diseases that putrefy the body live, and centipedes a foot long. This horrifying vision of danger and disease was conjured up by generations of explorers and writers about the Amazon but it lacks any real foundation in fact. Many of the diseases described in old travel books do not belong to the Amazon at all, but were imported from other continents. Generally though, the more horrific and colourful the travellers tales and the wider their readership, the more lucrative it was for the author and the publisher. Most Amazonian fauna are not very dangerous nor, on land, particularly large. I knew all this before I arrived in the Amazon, but when I came face to face with the first Bird-eating spider, *Avicularia avicularia*, the vision of a Green Hell came flooding back. I might have banished my preconceptions rationally, but clearly, had underestimated this disconcerting confrontation.

As we awoke the next morning, the forest was gloriously fresh, with hundreds of delicate shades of green gleaming in the sun and the uproar of the howler monkeys once more in the far distance. On this memorable day we came across a three-toed sloth, *Bradypus tridactylus*. Of all the animals in Amazonia, the sloth has to be the weirdest. Almost statuesque in their lethargy, gently stoned on the leaf buds they absent-mindedly chew all day, sloths exude astonishing calm amidst the bustle of the rainforest. They have long limbs, short bodies and a stumpy tail and the whole creature is covered in a pile of shaggy, grey, coarse fur and, often, green mould. Their faces seem to wear a permanent expression of sadness. They generally live an upside-down existence, hanging from trees secured by two or three (depending on the species) extremely strong claws. These claws look like hooks but could inflict serious wounds on the unwary. In Portuguese they are called *preguiça,* which means laziness. Considering their odd appearance and cautious, rather than slow, nature they are surprisingly good swimmers. The inhabitants of the Amazonas region, both Indians and descendants of the Portuguese alike consider the sloth a byword for laziness. It is very common for one native to call another, in reproaching him for idleness, *'bicho da Embaúba'* (beast of the Cecropia tree), the leaves of the Cecropia being the food of the sloth. From my diary again:

> "I am lathered in mud and sweat, one hand swinging a machete like a cutlass and the other groping to keep me upright on a rainforest pathway that clearly thinks otherwise. The narrow path crawls through the rainforest like a golden thread in a vast green tapestry, its twisted course dipping in and out of the oppressive verdant weft. Falling is a constant possibility as numerous high and narrow parts of the trail become slippery in the rainy season. The problem is that if you slip, you need to grab something to stop you falling, and if you do grab something, it can tear your hand. Scorpions, tarantulas, and lethal ants lurk in the leaf litter. Deadly vipers and fer-de-lances lie disguised as branches and roots, and even the flora threatens armed response. Thorns, hooks, and barbs tear at and shred clothes and skin, causing wounds that go septic in hours, and peaceful looking leaves cause cruel and unusual burns. You could also find yourself in dire need of re-hydration as heat and humidity raise the stakes while thunder rumbles in the distance like an empty threat. Travel is by river, but to enter the forest you have to wade through the current and climb the steep, muddy ridges except where the water runs too deep. We are soaked all day, changing into dry gear only at night and putting on damp clothes again in the morning."

"The wilderness throbs with life. From tiny poison arrow frogs to huge iridescent morpho butterflies, to toucans, parrots and hummingbirds no bigger than bees, the rainforest fauna comes dressed to impress. Howler monkeys, tamarins, and capuchins drop rotten fruit on you or worse while the three-toed sloth appears everlastingly still. Undisputed king of the rainforest is the jaguar. Seldom seen, rarely heard, it leaves only paw prints as it prowls the forest by night. It is at night that the sense of isolation is strongest in a forest too dense for satellite phones to work properly."

"Slowly the traveller comes to terms with the very idea of the river and at the end of each day, as I swung in my hammock with the gloomy forest all around me, I found myself being drawn deeper and deeper into the mysteries of this strange wilderness. After the cities and factories, villages, farms, woods and fields of the western European culture, it was a strange sensation adapting to a land dominated by two features only, water and forest."

"It sometimes seems the more you see of it the less real it appears. Land is not quite a correct description as the floor of the Amazon basin is so saturated with water that many explorers have perhaps more correctly described it as a river-land. The only roads are the occasional international highway like the Trans Amazonian link that is mostly a failure because it is impassible for more than half of the year during the rainy season. The rivers are the roads of the rainforest, 50,000 navigable miles of them are the trunk routes through the forest. Because the floor of the basin is so flat, large areas of forest are flooded every year, some even all year round, so most travel has to be on water. Often the riverbanks are drowned by ten to forty feet of water, and these floods extend from 25 to 60 miles inland on either side of the riverbed. Even when the rivers are not in flood, large expanses of water are left trapped in the undulations of the landscape. These are the *várzea* lakes in Brazil, and they intersperse the whole of the Amazon basin along the line of the rivers, looking from the air like dark commas in the forest."

"The Equator passes close to Belém through the mouth of the river and divides the whole region into two equal halves. To the north of it, the height of the rainy season is in June and to the south of it, in December or January. As the rivers rise and fall, first in the north then in the south, the whole Amazon pulsates with water like a valve in a giant pump. Perhaps one of the most astonishing discoveries about the waters of the Amazon is the way the river flows are affected by the torrents of the tributaries. The enormous pressure of the river Negro forces smaller tributaries to run backwards and a clear divide exists between the Negro with its black water and the Amazon with its coffee coloured water running side by side without mingling for mile after mile after their confluence."[63]

"Making progress along the waterways the surface was covered with a litter of broken trees, leaves, blossoms, small creatures, twigs, spume and dust. It looked like a marble floor that needed sweeping, faintly reflecting the overhanging forest as highly polished marble might do."

"The difficulty all the time was not being able to make valid comparisons in order to help understand the scale of the river. The forest, for example is the largest in the world, covering an area much greater than the size of France. Apart from the delta region, where grazing plains have been cleared, and a few dry areas of savannah in the lower Amazon, vegetation cloaks the entire Amazon basin, hanging right down to the water. Along all the riversides it continues creeping right up into the foothills of the Andes and overgrowing the sloping sides of the Guyana shield in the North and Brazilian shield in the South. It is at its most dense in the upper Amazon above Manaus. For hundreds of miles, it lines the riverbanks with a monotonous, sullen greenness, menacing and silent."

"The forest is the only one of its kind, a world complete in itself, living by its own laws, adapted to conditions that are, if not unique on our planet, at least matchless on this scale. There are no seasons as we know them, the rains come and go but the

[63] Goulding, M, Barthem, R, and Ferreira, E, *The Smithsonian Atlas of the Amazon.*

seasons are barely distinguishable in the upper Amazon, with the constant, humid, equatorial heat. At Manaus the temperature averages 81°F, ranging between the 70s and 90s all the year round, year after year. The flora and fauna have modified to these conditions and life goes on in what to us seems complete confusion. Shedding leaves, budding and flowering, moulting, pairing and breeding all occur at the same time. Every now and then after a heavy shower, plants burst into flower as if spring had arrived. In the middle of the day, the sounds of the animals are silenced, leaves droop and petals fall as if autumn had set in. From the almost bitter cold of the forest at night to the brightness of the morning to the languid heat of the afternoon, life seems to bypass a whole year's seasons. Yet, except for a few partially deciduous areas, the forest never turns brown. With all the heat and water vegetation grows with constant and excessive profligacy, as if sheltered by the cover of a giant conservatory."

"A strange and forbidding darkness is another feature that astonishes the traveller when encountering the forest for the first time. As the millions of trees wrestle each other for a share of the light, they grow to massive heights and extend their foliage like a green umbrella at the top, leaving the trunks below bare. Very little sunlight penetrates to the ground, only about 10 per cent where it occurs at all, and one walks through the forest in a gloominess with the lofty green vault of foliage far above. In its deepest retreats, the forest seems to be occupied in a savage but soundless conflict. Lianas, tropical climbing plants, wind themselves like boa constrictors round the tree trunks and arch themselves in great loops as they struggle upwards for a glimpse of light. The trees put out roots, which crawl on the ground like serpents, or line themselves with dominant flying buttresses. Epiphytes, parasitic air plants that live off the humid atmosphere or have their own water reservoirs, cluster close to the treetops."

"This enormous world of water and forest uninterrupted by the seasons and only dimly lit is strange enough to comprehend but this picture remains unfinished and still commonplace until another aspect is added: time. It is perhaps the Amazon's most astonishing attribute that it has remained comparatively unaffected and untouched for over 100 million years. To get an idea of what this means, one has to keep in mind that temperate forests, like those of Europe and North America, are post-ice age, a mere 11,000 years old, about one ten-thousandth of the Amazon's age. At the same time as parts of the Amazon forest were momentarily replaced with savannah during the drier climatic conditions that prevailed in the ice ages, many remote patches of rain forest survived this period untouched."

"The Amazon is thus one of the last places on earth where one can look deeply into the past, where one can touch a tree and become conscious that thin bark, buttressed roots, and its straight, light-seeking trunk are examples of plants still in their primitive state. As I became used to the Amazon I always had the strange feeling I was wandering the earth in primeval times, before the emergence of man. I left it with nothing but high regard for the resolve and skill, tenacity and indeed courage of Bates, the astonishing traveller and his contemporaries."

Part two

The Evolution of the Naturalist

Chapter 4

Plate 5.
Bates c1860

Plate 6.
Wallace c1845

With Wallace to the Amazons

In his last letter to Bates before their voyage, Wallace compares the holdings of beetles and butterflies in the *Muséum National d'Histoire Naturelle, (Jardin des Plantes)* in Paris with the collections in the insect room at the British Museum, concluding that both would benefit from the sort of venture they were planning. Wallace, however, aspires to go beyond the mere study of local collections:

> "I should like to take some one family to study thoroughly, principally with a view to the theory of the origin of species. By that means, I am strongly of [the] opinion that some definite results might be arrived at."[64]

[64] Beddall, Barbara G, *Wallace and Bates in the Tropics*, (London: Collier-Macmillan, 1969) & Wallace, *Alfred Russel, My Life, A Record of Events and Opinions*. 2 vols. (London: Chapman & Hall, 1906) A question always arises regarding which of the two of them first thought seriously about the origins Was it Bates or was it Wallace? Most authorities state that it was Wallace's idea and that he wrote to Bates about this well before they set off for the Amazons. Bates never claimed the initiative for himself. Bates's widow Sarah returned all Wallace's letters, but the one dealing with the origins has not since been found. Wallace, in *My Life*, remembered writing it sometime between 1845 and 1848, when they left for Brazil, but was not able to be more precise. There is no doubt it was discussed and it was one of their objectives. It is therefore reasonable to assume that they pursued this goal once there.

Clearly, Wallace was already thinking as a future biologist rather than a traditional collector. The decision to leave England for South America was certainly influenced by reading William Edwards's book, *A voyage up the River Amazon, including a Residence at Pará.* As Wallace explained:

> "This little book was so clearly and brightly written, described so well the beauty and the grandeur of tropical vegetation, and gave such a pleasing account of the people, while showing that expenses of living and of travelling were both moderate, that Bates and myself at once agreed that this was the place for us to go to if there was any chance of paying our expenses by the sale of our duplicate collections."[65]

The chance meeting with Edwards, who happened to be in London, settled matters. Edwards wrote letters of introduction to some friends in Belém (then known as Pará). Bates and Wallace had already contacted Edward Doubleday, curator of the British Museum's butterfly collections. Doubleday assured them that the whole of northern Brazil was comparatively unknown and the sale of a collection of original insects, land shells, birds, and mammals would easily pay their expenses. This must have pleased Bates and Wallace for, apart from their original shared £200, they had no financial resources until they sold some specimens.

The First Professional Naturalists

Bates and Wallace arranged a passage to Liverpool, saving money by travelling on top of the Liverpool stagecoach. They broke their journey at Chatsworth on 24th April 1848 hoping to see Joseph Paxton and to view the palm and orchid houses, then considered the finest in England. They wanted to identify the plants they should look for in the Amazon; however there is no record of their visit in Paxton's Chatsworth diary. On 10th April 1848, the Duke of Devonshire, worried by the possibility of a Chartist revolution had sent Paxton to London to report on developments and it was probable that he was still there.[66]

Once in Liverpool, Bates and Wallace contacted J. G. Smith who had collected butterflies in Pará state, on the Amazons, and at Pernambuco. The

---

[65] Clodd, Edward, Memoir of the Author, vii, xxi; and Wallace, *Alfred Russel, My Life, A Record of Events and Opinions.* 2 vols. (London: Chapman & Hall, 1906) 264.
[66] Chatsworth Archives, courtesy the Duchess of Devonshire.

night before their departure they dined with Smith and viewed his butterfly collection. Smith gave the two travellers more letters of introduction and useful information about (Pará) Belém and its environs.

There had already been two collecting expeditions to the Amazon, both large-scale and state sponsored. The Bavarians, Johann Baptist von Spix and Carl Frederick Philip von Martius, had collected in the Amazon between 1819 and 1820, spending some time at (Ega) Tefé, later one of Bates's favourite locations. Then between 1843 and 1847, Comte Francis de Castelnau, a correspondent of the *Muséum National d'Histoire Naturelle* in Paris, led an expedition through Brazil and the Andes.[67]

The Journey to Brazil

On 26th April 1848, Bates and Wallace embarked as the only passengers on the *Mischief*, a small trading barque of 190 tons. One month later, they were at anchor off Salinas in the estuary of the River Pará waiting for the pilot to take their vessel into the port. The sky was pure blue, the sea smooth and they were becalmed. As they approached Salinas, a great quantity of silt changed the colour of the water from Atlantic blue to Amazon coffee brown. According to Bates, the passage from Liverpool was swift and without incident, giving the travellers time to become accustomed to one another and their new environment, and to consider their future as they approached the coast of Brazil. Wallace said that the voyage was rough at times and, at one point, heavy seas wrecked part of the ship's bulwark. Perhaps Bates forgot this in his impatience to reach the tropics.[68]

Situated on the Pará River, Salinas was the pilot station for all vessels bound for Belém, the only gateway to the River Amazon. The ship lay six miles off shore in shallow water that made navigation difficult and resulted in a frustrating wait for Bates and Wallace who were eager to set foot on land. As the light faded, they peered at the distant shore through the Captain's eyeglass but saw only the faint silhouette of low-lying forests. The next day a pilot came on board and the *Mischief*, aided by a light breeze and the tide, reached Belém via the Baia de Marajó, passing the small fishing villages of Vigia and Colares. The passengers saw some native canoes although these seemed tiny when seen against the high walls of the

[67] Comte Francis de Castelnau was appointed French consul to Brazil in 1848.
[68] Wallace, A R, *My Life, A Record of Events and Opinions*. New edition. (London: Chapman & Hall, 1908) 145.

rainforest. The weather was hot and humid, the sky overcast and lightning streaked across the horizon.

Incidentally, this experience corresponded with my own. While I was on the Amazon, there was a thunderstorm almost every evening; the clouds sit down on the forest, first with furbelows spreading out, white, billowy and sedate, then they rush about with ragged fringes streaming behind and great columns rising high in the sky. In the forest there is tremendous excitement as every type of animal noise rises in a cacophony of sound. Clouds, some solid and others diaphanous, driven by powerful forces, move in different directions. There is a heavy cannonade for a few hours, and a great drumming in the river as the rain pounds down, peace returns and mist rises from the forest as the sky gradually clears while the animals fall silent once more.

Bates comments in *The Naturalist on the River Amazons* that the storm provided an appropriate greeting to his adventure. He and Wallace were astonished by the sheer size of the river. As their ship sailed close to one bank, the other was out of sight. At its mouth the Pará River is thirty-six miles in breadth and even at the city of Belém, it is still twenty miles wide. At that point, however, there are a series of islands, thus contracting the river view in front of the port.[69]

At dawn on 28th May, they finally stepped ashore at Belém. This first impression stayed with Bates for the rest of his life. In his book, he describes the clear blue sky, the low lying town, the white stucco houses with red tiled roofs, the crowns of palm trees and church spires alike towering above the skyline of the houses and, in the distance, the grey-green, low-lying wall of the forest. In the skies above, huge vultures wheeled round and round, floating on thermals and waiting for the opportunity to scavenge. Belém was noisy and colourful, as they had landed on one of the many Saints' days celebrated in this part of Brazil. Mr. Miller, the consignee of the vessel, met them and, according to custom, invited them to make use of his rocinha (small plantation) until they could find somewhere to live for themselves.

First Impressions

Once away from the port area, Bates observed the people and the place and was shocked by the pervasive poverty. The women's dress was colourful and

[69] Moon, H P, *Henry Walter Bates FRS, 1825-1892: Explorer, Scientist and Darwinian.* (Leicester: Leicester Museums, Art Galleries and Records Service. 1976) 25.

their industry seemed at odds with the indolence of the men who lounged about in the torrid heat. Bates noticed the differences between the races: the Negroes, descendants of the original slaves from West Africa, the coloureds with origins in the forests, and the whites of Portuguese birth. Although most appeared poor and underprivileged, the women especially caught his eye.[70]

Signs of neglect and decay were everywhere. The town appeared ramshackle and animals roamed wherever they wished, wandering in and out of gardens without hindrance. But 'amidst all, and compensating every defect, rose the overpowering beauty of the vegetation.' Belém was peaceful but presented an air of gentle dilapidation with the forest encroaching upon the houses. Of course, this meant that Bates and Wallace could find choice insect specimens close to their lodgings. The city had seen better days, public buildings were neglected and the roads and parks overgrown. During my own two visits, the city still had a similar graceful but neglected atmosphere.[71]

When describing the province of Belém and its inhabitants, Bates commented later that most of the tribes had become extinct or forgotten, at least those that had originally lived on the banks of the main river. Their descendants had amalgamated over time with the white and Negro immigrants. Mixed races now probably formed the greater part of the population and each group had a distinguishing name:

> "Mameluco denotes the offspring of White with Indian; Mulatto, that of White with Negro; Cafuzo, the mixture of the Indian and Negro; Curiboca, the cross between the Cafuzo and the Indian; Xibaro, that between the Cafuzo and Negro. These are seldom, however, well demarcated, and all shades of colour exist; the names are generally applied only approximately. The term Creole is confined to Negroes born in the country. The civilised Indian is called Tapuyo or Caboclo, (of European and Indian descent)."

The rainforest came right to the edge of the town and Bates and Wallace were amazed at the abundance of greenery and the magnificent tree

[70] Bates, Henry Walter, *The Naturalist on the River Amazons*. A Record of Adventures, Habits of Animals, Sketches of Brazilian and Indian Life, and Aspects of Nature under the Equator, during Eleven Years of Travel. Vol. 1 (London: John Murray, 1863) 7.

[71] In 1848 the population had fallen from 25,000 to only 15,000 because of the Cabañagem rebellion of 1819. Today the population is estimated to be 1,300,000.

specimens. Then they experienced their first tropical twilight on land accompanied by the cacophony of sounds that are so much part of the tropical scene. It rained and the refreshing cool air was welcome. Later into the night, as they became aware of the noises of the town nightlife mingling with the never-ending sounds of the forest, they climbed into their hammocks and slept soundly.

The first few days were occupied in unloading and unpacking their belongings and collecting equipment needed for their expeditions. Nets had to be assembled and more wooden collecting boxes were made to add to the existing stock. Guns were put together and primed ready for use and extra ammunition made up or acquired. They also purchased hammocks and household utensils. Everything had to be as portable and as jungle-proof as possible.

Bates and Wallace had been staying at Miller's house for about a fortnight when they heard of a similar but smaller house for rent in the village of Nazareth close to the centre of Belém. The house bordered the forest and was ideal for collecting expeditions. It belonged to a Portuguese tile manufacturer, named Danin, who owned a factory in the forest and lived nearby. The house, opposite the shrine to Our Lady of Nazareth, had an iron-grilled gate that led to the village green and was surrounded by thatched huts.

Bates explains that the virgin was a great favourite in Belém, credited with many miracles. Her image, a handsome doll about four feet high, with a silver crown and a garment of blue silk, studded with golden stars, was displayed on the altar of the shrine.[72]

Isidoro, a freed Negro slave, was hired as their cook and servant. Isidoro had previously worked for an Englishman but could not speak the language; he was to stay with Bates for many years. As their domestic situation stabilised, Bates and Wallace ventured further and further into the surrounding forest.

For a Victorian, Bates had little racial prejudice and this probably explains why he adapted so successfully to a country where it was essential for foreign travellers to assimilate with the local people. In an early letter home to his brother Frederick, he commented on the European traders' dress-coats, polished boots and top hats. He adopted the more practical dress of a coloured shirt, a pair of denim trousers, common boots, and old

[72] Bates, Henry Walter, *The Naturalist on the River Amazons* 1863 Vol 1. 86-94.

hat. Thus clothed, Bates, and Wallace who dressed similarly, roamed the forest lanes around the rocinha. During their first fortnight in Belém, they collected no less than four hundred different insects, many new to science.

Though captivated by the butterflies, Bates was disappointed not to see any large or spectacular animals. However, the travellers did see many new birds, especially parrots; these remained their favourites throughout their time in the rainforests. It was also disappointing that many of the small birds were much like those in England. However, small flocks of noisy and gaudy parakeets often appeared and proved a source of amusement. Once they had located their exact habitat, they found hummingbirds feeding in their hundreds on particular trees. In the skies above, the ever-present vultures circled.

Despite the absence of larger animals, Bates and Wallace saw lizards in abundance crawling over buildings and walls, and scurrying across pavements. Bates noted the geckos' speckled grey or ashy colours and the way their beautifully adapted feet enabled them to run over the smooth surfaces of ceilings and walls, as their toes expanded to form an adhesive cushion allowing them to move about securely.

Bates and Wallace resolved to walk through the forest to Senhor Danin's Mills. Initially they encountered a wall of forest rising up to 100 ft like a green cliff and festooned with various types of creepers new to them both. Here butterflies flew in large numbers, some reminding them of familiar British species, but even more glorious because of their size and colouring. For the first time they saw the *heliconiads* that were to play such an important part in Bates's life: he would eventually use these butterflies to illustrate mimicry.

After about a mile they found themselves for the first time in the primeval forests. The track led them to a place called Maranhäo. Bates was astonished by the immense size of the trees; some had circumferences of between twenty and thirty feet and trunks so tall that their lowest branches were one hundred feet from the ground. The forest canopy was like the arched roof of a cathedral with the columns of the trees rising with majestic arches to support it.

After a full morning's walk they reached the Mills and were courteously received by Danin who spoke English reasonably well. They completed the arrangements for the house they were to rent from Danin, discussed their expedition plans, and enquired about the general nature of the surrounding countryside. Danin's red-tiled house was painted white and raised on stilts above the flood line. It was located by the mill on the far bank of the Una near to its mouth where the river was about one hundred yards wide.

Danin had visited England where he was educating one of his two sons. As he introduced his associates each expressed a wish to travel to England or to send a son there. After this interesting and informative visit, Danin lent them a canoe and two Negroes to take them back down river to Belém.

Bates compared what he saw on his first forays into the tropical forests surrounding Belém and its gardens with familiar English species. He observed two swallow-tailed butterflies similar in colours to the English swallowtail, a white butterfly, similar to the English cabbage white, and two or three species of brimstone and orange coloured butterflies, which did not belong to the same genus as the English ones. Bates commented:

> "…a beautiful butterfly, with eye-like spots on its wings was common, the *Junonia lavinia*, and the only Amazonian species which is at all nearly related to our Vanessas, the (Red) Admiral and Peacock Butterflies."

Ants were everywhere and Bates wrote about them eloquently. He was amazed to see ants an inch and a quarter long and stout in proportion, marching in single file through the thickets. He watched the Saüba ants, *Atta cephalotes*, and marvelled at their dexterity, skill and determination as they struggled on the forest floor with their burden of leaf cuttings, a habit recorded in earlier books on natural history. As they march, their processions look like a multitude of animated leaves.

Every tree in the woods was festooned with various kinds of orchid, and a climber known as passiflora abounded, attracting curious butterflies in which Bates would become deeply interested. His first sight of the tiny long-tailed hummingbird with its emerald breast brilliant in the sun delighted him. Just to observe so marvellous a creature in its native habitat made the long sea voyage worthwhile.

Moonlit nights were particularly beautiful; the atmosphere was translucently clear and a light sea breeze produced an agreeable coolness. The rented house was close to the forest on three sides, so it was easy to enter the woods for a day at a time. A series of paths, simple to follow, brought Bates and Wallace to the house where Spix and Martius had stayed in 1819. Though now neglected and overgrown, the house conjured up visions of previous great explorations.

Bates was concerned that he was not seeing creatures in the profusion he had expected. The great variety of mammals, birds, and reptiles seemed to be widely scattered and excessively shy of man. The region is so extensive and uniform that animals are only occasionally seen in large numbers and then only in a very specific type of habitat.

Adjusting to a New Life

Bates and Wallace continued to be disappointed with what they saw in the rainforest until Bates looked up to the forest canopy and discovered the great array and diversity of creatures that lived there.

Both were surprised by the silence and gloom deep in the forest. Occasionally there were unexpected sounds and they were sporadically startled by a sudden shriek, yell, or scream that gave no clue to its origin. Perhaps some defenceless animal had met its fate at the jaws of a jaguar or from an Indian arrow. In the morning and again in the evening, as the howler monkeys started the uproar marking their territory, both men found the sound so fearful and harrowing that it lowered their spirits. My own experience was that the sound was so disconcerting it made the forest seem more wild, inhospitable and dangerous than it really was:

> "In the dull silences of the day or night there was often a tremendous crash as some great tree fell, forcing its way through the under-storeys of smaller trees as it dropped to the ground. The gap it created would soon experience rapid regeneration as other trees and plants took the opportunity to struggle ever upwards in the search for light and life." [73]

One of the letters of introduction provided by Edwards was to a Mr Leavens who subsequently invited Bates and Wallace to accompany him up the river Tocantins, a large tributary of the Amazon from the south. Leavens managed a rice mill and sawmill, owned by a Mr Upton and situated on a creek. Their first trip to the mills was by land and they took Isidoro with them:

> "At one time I had a Mameluco youth in my service, whose head was full of the legends and superstitions of the country. He always went

[73] Author's Amazon diary, *Sixty days on the river*, 2003-2005.

> with me into the forest; in fact, I could not get him to go alone, and whenever we heard any of the strange noises mentioned above he used to tremble with fear. He would crouch down behind me, and beg of me to turn back; his alarm ceasing only after he had made a charm to protect us from the *Curupira.* For this purpose, he took a young palm leaf, plaited it, and formed it into a ring, which he hung to a branch on our track."

After about a mile and a half they were in the primeval forest and by midday, after six hours hard walking, arrived at the mill on the banks of the Iritiri. Twelve miles overland from Belém, the creek was narrow and bounded on both sides by the loftiest of forests. There was a community of sorts close by that depended on trade with Leavens and the mill. The mills formed a large group of buildings, situated in a cleared tract of land, many acres in extent, though everywhere surrounded by uninterrupted forest. Leavens received them kindly, taking them to the best spots to find birds and insects. The mills were in an attractive setting, with a long meadow leading down to the stream. Cattle were kept on the meadow and were surrounded by a variety of water birds including many jacanas.

Bates and Wallace added considerably to their collections during these walks. Butterflies were plentiful, and during the course of a few days they saw and caught the sort of numbers that would have taken a whole summer to assemble in England. This was reassuring for they had to finance themselves by the sale of specimens. However, these acquisitions meant a huge amount of new work: daily the specimens had to be sorted, classified, set and stored in preparation for shipment.

In the Pará estuary Bates learned that the indigenous population originated from the Pernambuco area and differed from the Indians in the interior. These original inhabitants either were civilised, or had integrated with the white and Negro immigrants. Their distinguishing tribal names were forgotten and they now bore the general designation of Tapuyo, apparently one of the names of the ancient Tupinambás. The Brazilians call the still savage Indians of the interior, Indios or Gentios (Heathens). All the semi-civilised Tapuyos of the villages and the inhabitants of remote places in general speak the *Lingoa geral*, a language adapted by the Jesuit missionaries from the original idiom of the Tupinambás. On their second visit on foot to the mills, Bates and Wallace stayed ten days.

They explored the surrounding forests by canoe, the only means of transport, as the land everywhere was covered with impenetrable forests or water. All the settlements they encountered were on the waterside and the only communication was by river. Canoes would appear from nowhere paddled with great dexterity and skill, often by children.

Leavens wanted to go up river to see if he could verify reports that cedar grew in abundance between the lowermost cataract and the mouth of the Araguaia. A Portuguese trader had arrived with a quantity of worm-eaten cedar logs he had found among the floating timber in the current of the main stream. Bates and Wallace agreed to accompany Leavens but before leaving the mills, arranged further excursions with Leavens to the Tocantins as he was familiar with the languages of the area and adept in river navigation. Afterwards Bates and Wallace returned to Belém to send their collections to England, and prepare for their next journey to the interior.

Bates stayed in Belém for 18 months until 1849. During the first few weeks he witnessed the festivals in honour of the Virgin Mary that occupied much of the people's time.

Bates later explained that the festivals celebrated either anniversaries of events connected with individual saints or those in the life of Christ. Since independence, additional gala days commemorating national events had been added, but these too had a semi-religious character. The various holidays became so numerous and interfered so seriously with trade and industry that in about 1852, the Brazilian government obtained permission from Rome to abolish some minor festivals. Later still, many of the remaining festivals declined as populations became more mobile with the introduction of steamboats and railways. However, when Bates was in Belém at the end of the 1840s, the festivals were still in their full glory:

> "The women are always in great force, their luxuriant black hair decorated with jasmines, white orchids and other tropical flowers. They are dressed in their usual holiday attire, gauze chemises and black silk petticoats; their necks are adorned with links of gold beads, which when they are slaves are generally the property of their mistresses, who love thus to display their wealth…Numbers of young gaudily dressed Negresses line the path to the church doors with stands of liqueurs, sweetmeats, and cigarettes, which they sell to the outsiders."

Snakes abounded in the neighbourhood of Belém, particularly when it rained:

> "The majority of the snakes seen were innocuous. One day, however, I trod on the tail of a young serpent belonging to a very poisonous kind, the Jararaca (*Craspedocephalus atrox*). It turned round and bit my trousers; and a young Indian lad, who was behind me, dexterously cut it through with his knife before it had time to free itself. In some seasons snakes are very abundant, and it often struck me as strange that accidents did not occur more frequently than was the case."

It was the insects however, that occupied most of Bates's time and he quickly amassed a huge collection of butterflies from within an hour's walk of the town. The collection was ten times larger than the total number of butterflies found in Britain. He was astonished, just as I was, on first encountering the huge metallic blue *Morpho* butterfly with a six to seven inch wingspan. In my diary I noted:

> "As we drifted (in a canoe) *Morpho* butterflies, big as birds, flashed their electric-blue wings along the riverside, pugnacious insects, whirling in combat whenever male met male…A huge butterfly, the blue trembling on its wings like electricity, flew about our clearing. Sometimes it vanished, being hidden in its folded wings showing only the cryptic underside, hidden in the centre of some bushes."

Of these butterflies, Bates says:

> "When we first went to look at our new residence in Nazareth, a *Morpho menelaus*, one of the most beautiful kinds, was seen flapping its huge wings like a bird along the veranda. This species,(of butterfly) however, although much admired, looks dull in colour by the side of its congener, the *Morpho rhetenor*, whose wings, on the upper face, are of quite a dazzling lustre."

Beetles also demanded special attention. At first they seemed scarce. This apparent scarcity (observed by other travellers in the tropics) was attributed to the fact that the sun was so hot that beetles could not survive in exposed situations as they did in Europe. However, many hundreds of beetles of different families and species could be found in shady places.

To the Tocantins and Cametá in 1848

Confident in their new surroundings and travelling with Leavens, Bates and Wallace set off up the Rio Tocantins almost due south of Belém. The Tocantins is the third longest of the Amazon tributaries. It runs for 1,600 miles so this expedition must be regarded as a major venture into the interior. Wisely, Leavens spent a considerable time before they set off helping the two young travellers with detailed preparations.

This part of the river system is close to the sea, and the Rio Tocantins is tidal; it is so wide that it is often difficult to see the far bank. Leavens therefore hired a two-master sea-going v*igilenga*, twenty-seven feet long, with a flat prow and great breadth of beam and fitted for use in heavy seas. The boat had no deck but there were two arched wickerwork awnings, thatched with palm fronds to provide shade. The travellers anticipated being away for about three months so had to take substantial provisions. Quantities of farinha, dried fish and rum rations for the crew, and luxuries such as sugar, rice, coffee and tea were stowed on board. Other stores included guns, ammunition, and equipment for collecting and preserving specimens, together with plenty of copper coin for establishing good will with the indigenous people. They obtained safe passages for the journey and set about finding a crew. Leavens made most of the arrangements and provided two crewmembers from his workforce at the mill. Bates and Wallace took Isidoro, by now an invaluable companion, and recruited a young boy called Antonio, one of many who had attached themselves to the two men in Nazareth.

The principal crewmember was Alexandro, one of Leavens's men, an intelligent and well-disposed young Tapuyo, an expert sailor and hunter. Being a native of a district near the capital, Alexandro was a civilised Tapuyo, a citizen as free as his white neighbours were. He spoke only Portuguese. Three years later Bates met him again in Belém wearing the uniform of the National Guard, and Alexandro called on Bates often to talk about old times.

On 16th August Bates writes enthusiastically to his brother Frederick saying they had caught 460 different species of butterfly in the forests around the mills. On 26th August, they finally sailed.

They rested on the floor of their 'cabin', and waking refreshed from a good night's sleep found the boat drifting with the tide along a wall of impenetrable vegetation plunging down to the water. This never-ending wall of forest was to persist until 28th August. The next day they entered the

Moju, a stream slightly inferior to the Thames in size, connected about twenty miles from its mouth by means of a short, artificial canal with a small stream, the Igarapé-mirim, which flows the opposite way into the water-system of the Tocantins.

Later that day they reached the Tocantins, one of their Indians announcing *'La esta o Parana-uassu!'*, (Behold the great river). The trees had changed and now included a high proportion of palms. They anchored and went ashore to prepare food and drink. While this was going on, Bates explored inland and was amazed to discover that the luxuriant palms were almost all of one species, the gigantic fan-leaved *Mauritia flexuosa.*

They reached Cametá on the morning of 30th August; there some hands deserted, preferring the pleasures of drinking on shore to crewing the boat. The sailing party moved on quickly to Vista Alegre, fifteen miles above Cametá, in the company of Senhor Laroque, an intelligent Portuguese merchant. Laroque took them to the residence of Senhor Antonio Ferreira Gomez, from whom they hoped to secure more crew. This was the first up-country estate Bates and Wallace had visited:

> "....a fair sample of a Brazilian planter's establishment in this part of the country. The buildings covered a wide space, the dwelling house being separated from the place of business, and as both were built on low, flooded ground, the communication between the two was by means of a long wooden bridge. From the office and visitors' apartments a wooden pier extended into the river. The whole was raised on piles above the high-water mark. There was a rude mill for grinding sugar cane, worked by bullocks; but cashaca, or rum, was the only article manufactured from the juice. Behind the buildings was a small piece of ground cleared from the forest, and planted with fruit trees - orange, lemon, genipapa, goyava, and others; and beyond this, a broad path through a neglected plantation of coffee and cacao, led to several large sheds, where the farinha, or mandioca meal, was manufactured. The plantations of mandioca are always scattered about in the forest, some of them being on islands in the middle of the river. Land being plentiful, and the plough, as well as, indeed, nearly all other agricultural implements, unknown, the same ground is not planted three years together; but a new piece of forest is

cleared every alternate year, and the old clearing suffered to relapse into jungle." [74]

They paused here for two days in the forest collecting birds and insects, discovering new species they had not found at Belém. They also lost another crewmember when Antonio deserted. The crew was so depleted they considered returning to Belém for sheer lack of hands. At the last moment Senhor Gomez lent them two slaves to accompany them as far as Baião.

Considering this background, and in particular Wallace's Owenite leanings, it is surprising they said little about slaves. Bates seems to have considered them simply as additional hands necessary to ensure the journey could continue. By 2nd September they were about 25 miles from Vista Alegre towards Baião, passing between islands on which there were many houses. In low situations these had an unfinished appearance, mere frameworks raised high on wooden piles and thatched with the leaves of the Ubuçu palm. For the next stage of the journey, from Vista Alegre to Baião, all on board were obliged to row the craft; they had to stand on a raised deck, formed by a few rough planks placed over the arched covering in the fore part of the vessel, and pull with all their might with their backs to the stern. Early on the morning of 3rd September, they arrived at Baião, built on a very high bank, with a population of about 400 inhabitants. Senhor Seixas had given orders for rooms to be prepared for them as he was away at his sitio, (establishment) and would not be here until the next day.

Wallace highlights the importance of letters of introduction by quoting in full a letter they carried from Senhor José Antonio Correia Seixas & Co., Baião:

> "Friends and gentlemen, knowing that it is always agreeable for you to have an opportunity of showing your hospitable and generous feelings towards strangers in general, and more particularly to those who visit our country for the purpose of making discoveries and extending the sphere of their knowledge; I do not hesitate to take advantage of the opportunity which the journey of Mr. Charles Leavens and his two worthy companions presents, to recommend them to your friendship and protection in the scientific enterprise which they have undertaken, in order to obtain those natural productions which render our province a classic land in the history of

[74] Woodcock, George, *Henry Walter Bates: Naturalist of the Amazons.* (London: Faber & Faber, 1969) 43.

> animals and plants. In this laborious enterprise, which the illustrious (elites) travellers have undertaken, I much wish that they may find in you all that the limited resources of the place allows, not only that whatever difficulties they encounter may be removed, but that you may render less irksome the labours and privations they must necessarily endure; and for men like them, devoted to science, and whose very aliment is Natural History, in a country like ours abounding in the most exquisite productions, it is easy to find means to gratify them. I therefore hope, and above all pray you to fulfill my wishes in the attentions you pay to Senhor Leavens and his companions, and thus give me another proof of your esteem and friendship. Your friend and obedient servant, João Augusto Correia."[75]

The unloading of apparatus and store-boxes from the canoe may seem simple, but was actually complex and exhausting work. Rapid movement produces excessive perspiration in this climate and physical exertion of any kind can be exhausting, even for fit young men. Essential pieces of luggage included the guns and other collecting tools, dissecting instruments, and everything needed for preserving and storing specimens.

Insects were caught by the aid of muslin butterfly nets and sugared traps, but lighted moth traps do not seem to have been part of their equipment, nor were they much used by other collectors at this time. 99 Killing-bottles were introduced around 1845 and the newly discovered chloroform was soon used in them. This discovery came too late for Bates's and Wallace's collecting but they may have used this method in later years. From 1845 onwards, the use of potassium cyanide in killing bottles became more common. This method of killing insects quickly and cleanly was a welcome alternative to crushing them on the thorax or pinning them alive. Small specimens of all kinds and the minute details of larger ones could be examined with a magnifying glass, but most naturalists now managed to include a microscope in their baggage. Although the microscope had been invented in the sixteenth century, such instruments were too rare and valuable to risk on expeditions. In the 1830s, however, they became cheaper and more readily available, costing between thirty shillings and four guineas, and it is probable Bates and Wallace had one. In addition to various nets, apparatus for

[75] Wallace, A R, Travels on the Amazon and Rio Negro, (London: Ward, Lock & Co Ltd) 1853. 41.

collecting insects would also have included pins of all sizes, small pillboxes for live specimens such as beetles, and wooden store boxes lined with cork for storing dead insects.

Each collector needed at least two guns, as well as the equipment required for carrying and occasionally making ammunition. Shotguns rather than live traps were used to collect most small animals, and local indigenous people would have brought in a variety of specimens they had trapped or killed with arrows or spears. Detonators, first patented in 1807, were adopted early in the nineteenth century, replacing the flint and spark. Around 1850, breech-loading guns and, later still, self-contained cartridges were introduced to England from France. These inventions gradually replaced the old muzzle-loaders and greatly improved the collector's chance of success. While small specimens, such as insects and spiders, could be preserved simply, mounted on pins and stored in pest-proof boxes, birds and mammals were usually dissected, skinned and cured with arsenic salts. Specimens of intermediate size, and soft-bodied or aquatic species had to be preserved intact. The excise duties imposed on glass in England during the Napoleonic Wars were removed in 1845 and this led to an abundance of cheap glass storage jars and bottles with cork bungs. Specimens were placed in these and spirit added to prevent decay. Local spirit was commonly used and specimens from the West Indies and South America often arrived in England preserved in rum. Bates and Wallace must have had good equipment because their specimens were of the highest quality when they reached Stevens's saleroom in London. This could not have been achieved had they stinted on collecting methods or equipment. Transporting set insects was even more difficult. Before leaving for the Amazons, Wallace explains how he and Bates intended to deal with this problem. They had gone to the India Museum in London to see a Mr. Horsfield who had collected in the subcontinent and developed a method for transporting delicate entomological specimens by sea.[76]

Bates was amused, though probably rather annoyed, by the laid-back attitude of the people encountered at Baião, and commented that it took time for the European to adjust to such an unhurried style of living. Once established in their temporary lodgings Bates and Wallace were surrounded by a crowd of curious loungers. One of them, a Mameluco named Soares, an Escrivão or public clerk, invited Bates to view his home and library. Bates was surprised to find a number of well-thumbed Latin classics: Virgil,

[76] Extracts from the Correspondence of Mr. H. W. Bates. (1855) *Zoologist* 13: 4549-4550.

Terence, Cicero's Epistles, and Livy. He was not yet sufficiently fluent in Portuguese to converse freely with Soares to ascertain what use he made of these books. It was an unexpected sight, a classical library in a mud-plastered and palm-thatched hut on the banks of the Tocantins in the depths of the Amazon rainforest. Later, during his stay on the Amazon, Bates became fluent in both Portuguese and German. He could also make himself understood in the *Lingua geral* of the Indians, however he did not return to see Soares again.[77]

Bates and Wallace enjoyed visiting this first typical Amazonian village. They liked the scenery and enjoyed the climate which was drier than in Belém. In the area near to Baião the forest was secondary; because of the number of settlements, the ground had been cleared of trees and cultivated but then the forest re-generated when cultivation ceased. Bates observed the mix of races but thought they had little enthusiasm to make a good living, leaving their plantations in semi-decay. They were courteous but lazy. When asked why, they said that it was useless trying to plant anything as the Saüba ant (leaf cutter) devoured the young coffee trees; anyone attempting to beat the ants was bound to fail, so planting was futile.

Senhor Seixas provided them with two more crew, killed an ox in their honour and was generally helpful. He did not however introduce his family and Bates and Wallace thought this unusual. Bates comments that he caught a glimpse of the pretty wife who was a Mameluco woman. She had a little girl with her. Both wore long dressing gowns made of bright-coloured calico print, and had long wooden tobacco-pipes in their mouths. Bates and Wallace stayed at this place but were not comfortable as the room where they slept and worked had formerly served as a storeroom for cacao, and at night they were kept awake for hours by rats and cockroaches, which swarm in all such places. The latter were running about all over the walls; now and then one would come suddenly with a whir full into Bates's face, and get under his shirt if he attempted to jerk it off. As to the rats, they were chasing one another by the dozens all night long over the floor, up and down the edges of the doors, and along the rafters of the open roof.[78]

Their stay was brief and, on 7th September 1848, they sailed at dawn from Baião, Senhor Seixas having generously provided them with fresh

---

[77] M Lee, 1985: 11.

[78] Clodd, Edward, Memoir in Unabridged (1863) commemorative Edition of *The Naturalist on the River Amazons: A record of Adventures, Habits of Animals, Sketches of Brazilian and Indian Life, and Aspects Nature under the Equator, during Eleven Years Of Travel.* (London: John Murray, 1892) 65.

provisions. The water was shallow and the crew were able to shoot fish with a bow and arrow, a technique Bates had not seen before.[79]

For a while, the boat glided through channels between islands with long, white, sandy beaches, over which ran occasional aquatic and wading birds. They went ashore frequently and for the first time Bates saw the tracks of a jaguar, later finding a rich haul of turtle eggs to supplement their rations. He also encountered Caiman for the first time, a large and ugly specimen that reared its head and shoulders above the water just after he had bathed nearby.

On 9th September, they reached a small cattle estate but moved on after failing to recruit more crew. By 10th September they had reached Patos, a small settlement of no more than a dozen houses, where Leavens attempted to recruit men to accompany them to the Araguaia. No amount of money would induce anyone to agree - not because of fear of the unknown but because of a general lethargy. There may have been other reasons for the reluctance, unknown to Bates and Wallace at this stage. There were dangerous cataracts between Patos and Araguaia, the most notorious of which, *The Inferno*, had claimed the lives of many local men.

The further they travelled the more elusive Leavens's cedar trees became, despite the quantity seen floating down the river. Reports of cedars were always vague and enticing; they were told the tree was plentiful somewhere, but no one could say precisely where. Bates believed that cedars grew like all other forest trees, in a scattered way and not in great stands. Leavens was told that there were cedar trees at Trocara, near some fine rounded hills covered with forest visible from Patos, so they searched there. Several families lived at Patos, and Bates once again commented on the charms of the younger women:

> "Women, old and young, some of the latter very good-looking, and a large number of children, besides pet animals, enlivened the encampment. They were all half-breeds, simple, well-disposed people, and explained to us that they were inhabitants of Cametá, who had come thus far, eighty miles, to spend the summer months."

A native they encountered volunteered to take Bates, Wallace, and Leavens into the forest to show them cedar trees. On this walk they saw for the first time the splendid Hyacinthine macaw *(Macrocercus hyacinthinus,* Lath.

---

[79] *The Naturalist on the River Amazons*, Bates (1863) Clodd, Edward (1892): 65.

the *Araruna* to the indigenous people), one of the finest and rarest species of the Parrot family. This macaw is over three feet in length with splendid purple-blue feathers; it flies in pairs and feeds on the hard nuts of several palms, but especially the Mucujá *(Acrocomia lasiospatha)*. Bates watched it feeding on nuts which he found almost too hard to break using a hammer and all his strength, but which the parrot crushed effortlessly to a pulp in its beak. As a result of over hunting, this magnificent bird no longer flies near Patos. However in 2003 I saw several pairs further to the south in the Mato Grosso do Sul.

Leavens was indifferent to the languid attitude of the people of Patos and its officials, the inspector, constable, and the governor, all of whom he thought slippery customers. This was crucial because the officials may have dissuaded the few men who might have become crewmembers. Bates thought the people idle and useless but partly excused them as a festival was taking place. The men in particular were often drunk with caxiri, an Indian drink made by soaking mandioca cakes in water until they ferment, producing a taste something akin to new beer.

As they were unable to secure more recruits, Leavens abandoned his original intention of going further upstream in search of the cedar, but he was a considerate man and recognising the enthusiasm of the two young explorers, agreed to travel upstream with them to the cataracts near Arroios. This gave the expedition more impetus as there was now a clear and attainable objective. As they progressed, the river became more picturesque and the shores were fringed with beaches of glistening white sand. On 14 and 15th September, they stopped several times to explore ashore, making their longest excursion to a large shallow lagoon, congested with aquatic plants, which lay about two miles across the campo. At a place called Juquerapua, they engaged a pilot to conduct them to Arroios. A few miles above the pilot's house, they arrived at a point beyond which their large canoe could go no further; for they were approaching the rapids.

On 16th September they boarded a large *montaria* lent to them by Senhor Seixas. The *vigilenga* was anchored close to the rocky islet of Santa Anna, to await their return. Isidoro was left in charge with their mulatto José, who had fallen ill since leaving Baião. Those remaining with Bates and Wallace were Alexandro, Manoel, and the pilot, a sturdy Tapuyo named Joaquim: *'scarcely,'* Bates concludes, *'sufficient crew to paddle against the strong currents'*. By mid morning they arrived at the first of the rapids called Tapaiunaquara. The river, which was about a mile wide, was choked up with rocks, a broken ridge passing completely across it. Between these confused piles of stone the currents were fearfully strong, and formed numerous eddies and whirlpools.

After ten hour's hard work they arrived at Arroios at about four o'clock in the afternoon, but found it little different from the other settlements - a few houses built on a high bank. Arroios was a stopping station for men going to the mines in the interior and for those making their way to the falls. Bates admired the way the small crew battled with the strong currents. Although more than a mile wide, the river was strewn with huge boulders with water rushing between. The main waterfall was about a quarter of a mile wide and Bates, Wallace and Leavens climbed to a high point to get the best view. They concluded that with good luck, a small craft and skilled oarsmen would get them through. They were impressed by the wildness of the scene as they viewed the forest that stretched for range after range over wooded hills.

On returning to the place where they had left their *vigilenga* they found poor José, the mulatto, much worse, so they proceeded to Juquerapua to procure aid. José was eventually cured by local herbal medicines but this took time and Bates and Wallace accepted the hospitality of Senhor Joaquim while they waited.

As a change from the confinement of their canoe Bates slept ashore, having obtained permission from Senhor Joaquim to sling his hammock under his roof. The house, like all others in these remote parts of the country, was a large open, palm-thatched shed, having one end enclosed by means of partitions also made of palm-leaves, so as to form a private apartment. Under the shed were placed all the household utensils, earthenware jars, pots, and kettles, hunting and fishing implements, paddles, bows and arrows, harpoons, and so forth, all the paraphernalia of domestic living. One or two common wooden chests served as containers for the women's clothing. There was no other furniture except a few stools and the hammock, which answered the purposes of chair and sofa.

There was much formality in the social life of these half-wild mamelucos, which Bates believed was chiefly inherited from their Indian ancestors, with some copied from the Portuguese.

There were many other strangers under Senhor Joaquim's roof besides Bates, Wallace and Leavens, among them mulattos, (mixed blood female African ex-slaves/male Portuguese), mamelucos, (mixed blood female Amerindian/male Portuguese), and pure Amerindians. Houses only occur at rare intervals in this wild country, and hospitality is still freely given to the infrequent passing traveller. After a frugal supper, a large wood fire was lit in the middle of the shed, and all then turned into their hammocks and began to talk. On this occasion, a few fell asleep but others told stories for several hours. Some related their adventures while hunting or fishing; others recounted myths about the *Curupira*, and other demons and spirits of the forest.

Later, as they descended the river, Bates and Wallace stopped frequently and added to their collections. Two month's work had secured over 400 species of butterflies, 450 species of beetles, the same number of other insects, and many bird specimens. They now sailed down the lower part of the river by a different channel to the one they ascended by, and frequently went ashore on the low islands in mid-river. They discovered that the islands were the habitat of India-rubber trees and found several people encamped collecting and preparing the latex.

They had reached a point opposite Cametá by 24th September, and found the islands planted with cacao, the tree that yields the chocolate nut. The forest is not cleared for this purpose, but the cacao plants are randomly planted amongst the trees. There were many houses on the banks, all raised above the swampy soil on wooden piles and furnished with broad ladders for access.

By 26th September, they were clear of the islands, and moved into the sea-like expanse of waters forming the mouth of the Tocantins. The river was now at its lowest level but still a mile and a quarter wide and fresh-water dolphins rolled in shallow places. Dolphins are more familiar as marine mammals although several species are found in the larger rivers of South America. Bates describes two species of dolphin from the Tocantins and knew of a third species on the upper Amazon. On 30th September, after threading through the labyrinth of channels linking the Tocantins and the Moju, they arrived back at Belém.

It was now October 1848. Hitherto, the expedition had been a joint venture with Bates and Wallace acting together with a common interest as well as purpose. Now in what seems a sudden decision, they decided to part, travel separately, and collect individually. It is possible that the parting was due to some argument or dispute, but extant letters and memoirs provide few clues as to the reason for it. The fact that prizes such as rare or beautiful specimens of butterflies or beetles could not be shared between two great enthusiasts may have been part of the problem. When a fellow collector – even a close friend or colleague – finds a better or more desirable object, complicated emotions can arise. There can be admiration, envy, a sense of anxiety, distress, and feelings of inferiority or, in extreme cases, rage –

perhaps a combination of all. Vladimir Nabokov sums up the entomologist's passion and dilemma:

> "Few things indeed have I known in the way of emotion or appetite, ambition or achievement, that could surpass in richness and strength the excitement of entomological exploration. From the very first it had a great many inter-linking facets. One of them was the acute desire to be alone, since any companion, no matter how quiet, interfered with the concentrated enjoyment of my mania. Its gratification admitted of no compromise or exception."[80]

Bates and Wallace were not to work together again except during a brief meeting in Manaus in 1850, but there was no sign of acrimony in their later correspondence. By the time they parted both had become more capable of dealing with the difficult and demanding tropical environment, but the decision to part can hardly have been taken lightly; travelling alone only accompanied by local native helpers was difficult, lonely, and dangerous at times.

The conclusion must be however that they did quarrel. In a letter to William Hooker, the explorer Richard Spruce, who arrived in the Amazon in the summer of 1849 accompanied by Wallace's brother, confirmed that a serious disagreement had occurred between the two men who completely avoided each other for more than a year. William Hooker must have asked Spruce to look out for Bates and Wallace, for Spruce wrote:

> "I forget to mention that we have several times seen Mr Wallace. He and Bates quarrelled and separated long ago."[81]

There is even an interesting remark by Bates in a letter to Stevens from Belém in August 1849 that suggests that he would wish not to be alone. Wallace and his brother Herbert had departed up river three weeks previously for Manaus and the Rio Negro. Bates writes:

> "I should have liked a sympathising companion better than being alone, but that in this barbarous country is not to be had. I have got a

---

[80] Nabokov, Vladimir. *Speak Memory, An Autobiography evisited,* (London: Penguin, 1967) 99. Also Johnson, Kurt, & Coates, Steve, *Nabokov's Blues: The Scientific Odyssey of a Literary Genius.* (New York. McGraw-Hill, 1999) 31-33.

[81] Richard Spruce to William Hooker, 3rd August 1849, Letters from Spruce (1842-1890), no. 259, Archives of the Royal Botanic Gardens, Kew.

half-wild coloured youth, who is an excellent entomologist, and have clothed him with the intention of taking him with me as assistant."[82]

Much later in life Wallace accepted an invitation to lecture on zoogeography before the Royal Geographical Society where his old colleague was the assistant secretary. However the formal tone of his letter addressed not to *Dear Bates* but to *Dear Sir*, underscores a drifting apart between the two men.[83]

After Wallace's departure, Bates became increasingly concerned with social issues. He noted that, when Wallace and he had arrived, Belém had not fully recovered from the effects of a series of revolutions, ultimately the result of the extreme animosity between the native Brazilians and the Portuguese. The native Brazilians had finally appealed for the support of the Indian and mixed coloured population.

At this time Bates sets down for us the daily routine of a naturalist in the tropics:

"...We now settled ourselves for a few months regular work... It was the end of the wet season; most species of birds had finished moulting, and every day the insects increased in number and variety...We used to rise soon after dawn, when Isidoro would go down to the city, after supplying us with a cup of coffee, to purchase the fresh provisions for the day. The two hours before breakfast were devoted to ornithology...After breakfast we devoted the hours from 10 a.m. to 2 or 3 p.m. to entomology; the best time for insects in the forest being a little before the greatest heat of the day...Our evenings were generally fully employed preserving our collections, and making notes. We dined at four, and took tea about seven o'clock. Sometimes we walked to the city to see Brazilian life or enjoy the pleasures of European and American society. And so the time passed..."

Picture Bates hanging bird skins to dry in paper cones attached to a line strung near the ceiling, having removed the flesh and bones for examination and dissection. The contraption was to keep specimens out of the reach of rats, which nevertheless sometimes managed to walk skillfully along the lines

[82] Mr. H. W. Bates. *Zoologist* 8: 2667.

[83] Letter from Wallace to Bates, 10th November 1876. Correspondence Block 1871-1880, Archives of the Royal Geographical Society, London.

and eat an accessible foot or wingtip. Isidoro is skinning a bird, not an easy task. Beside him are scissors, knives, nippers, forceps, and a pepperbox of pounded chalk, a jar of arsenical soap, needles and thread and the cotton wool to fill the hollow bird-skin. At another time, he is cleaning land-shells. After killing and removing the creature inside, he replaces the operculum and packs the shells carefully in boxes. Now Bates is laying out the botanical specimens brought in on this particular day, pressing them between sheets of coarse brown paper, and placing the pile between boards weighed down with stones. He must record facts observed in the morning's explorations and make sketches of the forms and colours that death would destroy. He would immerse beetles in boiling water and then in spirits or even rum, kill Lepidoptera with the vapour of prussic acid, pinning them through and fastening down their wings on special boards with paper or card braces, finally placing them in the setting boxes.

Chapter 5

Caripi and the bay of Marajó

Bates's and Wallace's friendship bore the hallmarks of an intense relationship until in the Amazons they realised that each was not quite the person they thought. There are unspoken distributions of power in all relationships and even though they shared the experience of similar backgrounds Bates was without doubt the dominant partner, and in the early years, his knowledge of field natural history was superior to Wallace's.

After just six months living and working together their feelings for each other were probably wearing thin. Wallace was almost certainly moody and Bates most likely could not back off under those conditions and give Wallace more space. In many ways, they depended on each other too much for this to happen and Bates probably closed in on Wallace because that was his nature. Perhaps then, they got on each other's nerves and differences in their relationship hardened. Arguments over silly incidental things could have followed reinforcing the differences. The problem was also probably one of boredom.

When they first met Bates and Wallace would have behaved in the normal manner, seeking out cohorts whose competence matched their own. For an expedition to the Amazon that would have included self-reliance. In friendship that could have included humour but whatever, it was intended to reflect their own sense of identity and reinforce it. However sometimes, the novelty can and does wear off. Difficulties arise when people are over exposed to each other and fail to react to external conditions in a way that influences subsequent development. Soon Bates and Wallace would have been able to instinctively guess what the other was going to say or do and there would then have been less reason to try to impress.

As partners, they had set out to solve the riddle of the origin of species. Wallace would have probably dominated the development of these ideas although Bates would have remained the prevailing partner in the relationship. The inevitable result would have been a power imbalance. Arguments would have become one sided and discussions could have become contentious.

At some point Wallace decided to move out, but he never divulged the reasons why he did so. Perhaps he and Bates had simply got too close to each other and the only option for sanity was to put some distance between them. This is not a decision that everyone is wise enough to make but in this case it ensured they were to remain friends for the remainder of their lives, but never again as close as during those early days on the Amazon

Wallace relocated to the house of the Swiss consul in Belém, a Monsieur Borlaz, and stayed with him for a month. Then on 3rd November 1848 he set off with a man called Yates, who was an orchid collector, on a three-month collecting trip to the Island of Marajó leaving Bates behind.

Meanwhile, Bates was mixing with the local people and his circle of friends began to resemble the multi-racial nature of Belém itself. He wrote to Stevens:

> "I get on well with the Indians being far more at home and friendly with them than with the Brazilian and European residents. The English people here, you will be sorry to hear, have not shown a disposition to assist us in the least all along."[84]

Bates hired an elderly Negress called Aunt Rufina who cared for him and most of the time he held court with the Brazilian community of Belém. The Europeans seem to have cold-shouldered him; the British mercantile community and colonial style society, which maintained their precarious social standing by sticking to the formal rules of their culture, deplored seeing an Englishman going native. Bates's casual dress alone would have caused comment from the top-hatted European gentlemen who saw him wandering the streets with his butterfly net, in his check shirt, denim trousers and old straw hat, whilst communing with the local people.

Bates wrote to his brother Frederick:

> "I have collected every day a splendid box of butterflies besides other things, always taking something new; and, in spite of the furious heat of the sun and great fatigue, enjoyed myself amazingly."[85]

Social Understanding

Although nominally a Unitarian, Bates was influenced by the anticlerical spirit of the times and professed to be an agnostic like Wallace. This would have predisposed him to have a constructive attitude towards the people he met in the forests. He would have been positive, as Unitarians believed that

---

[84] Letter to Stevens dated 10th March from Santarém: *The Zoologist*, 11 (1853) 4113-4117.
[85] M Lee, 1985: 26.

people are worth caring about whoever they are. Bates would have supported the idea of equality for all and would have been concerned for the beliefs of these individuals. However, although he found them full of novelty, he reacted ambivalently towards their celebrations and found them wanting. He thought the uneducated Portuguese immigrants irrationally superstitious and in the Negroes found a purer devotional feeling towards these events. Bates thought the Indians, despite having their own patron, São Tomé, had little religious belief or sentiment at all.[86]

He remained on the Lower Amazons for almost a year after his return from the Tocantins, before departing on his first journey to the upper river on 6th September 1849. During this year he left Belém on two occasions, once to spend several weeks at the hamlet of Caripi (on the island of Carnapijó) on the eastern shore of the Bay of Marajó, and later to re-visit the Tocantins, where he made the modest town of Cametá the centre of his collecting.

He obtained permission to spend two or three months at Caripi which was about twenty-three miles from Belém, and gained a reputation for the number and beauty of the birds and insects found there. It was located at the northern end of the Ilha das Onças (Isle of Tigers), which faces the city of Belém, and Bates bargained for a passage with the Cabo of a small trading-vessel, which started its journey on 7th December 1848.

Aboard the boat with Bates and Isidoro were eleven people: the pilot, his pretty mulatto mistress, five Indian canoe men, three young mamelucos who were tailor-apprentices, and a heavily chained runaway slave (slavery still being the order of the day in Brazil until the 1880s). The vessel took him to a red-tiled mansion located in the forest on the shores of a small bay at Caripi. This was a run down hacienda owned by Archibald Campbell.[87]

---

86 Woodcock, George, *Henry Walter Bates: Naturalist of the Amazons*. (London: Faber & Faber, 1969) 69.

87 From the time the Portuguese formally acquired Brazil from the Dutch in 1661, they used slaves. At first, the local Indian population was employed, but the Amerindians, whom the settlers referred to disparagingly as *burgres* (buggers), were unable to withstand enslavement; they died rapidly from smallpox, measles, respiratory infections and venereal diseases. The only other source of labour was Africa.

The facts of Brazilian slavery have been notoriously difficult to come by, for in 1891 all the records of slavery, the log books of slave ships, custom house records, documents of sale, ownership papers, were ordered to be destroyed to prevent any stigma attaching to the families of former slaves. However, in 1825, Alexander von Humboldt, set Brazil's slave population at nearly two million, out of four million. In 1830, the English abolitionist, Thomas Fowell Buxton, estimated imports at 100,000 a year, but recent estimates place the average import of slaves in the 1830s at about 34,000 a year.

In Brazil, as in other Roman Catholic nations, slaves were overall better treated than in Protestant countries. There were a number of reasons for this. For one thing, Spain, Portugal and France all had detailed laws concerning the treatment of slaves. For another, the Church took a close interest in the lot of the slaves, encouraging church marriages and opposing the separation of families. Thirdly, the Portuguese, like the Spanish, had little race or colour prejudice. In the British West Indies and in the southern United States, the child of a slave woman by her owner

Campbell had been hospitable to Edwards during his travels and had recently helped Wallace. Bates asked if he could stay there for two months and Campbell readily agreed. Campbell had a reputation for helping travelling naturalists and being genuinely interested in their work. Bates took over the rooms he needed in the house which he found large and substantially built. It was here that he spent his first Christmas day in a foreign land. This was celebrated as men, women, and children were busy in the chapel all day on 24th December decorating the altar with flowers and strewing the floor with orange leaves. As there was no priest available, Mass was dispensed with and the service consisted of a long litany and a few hymns that enchanted Bates. A small image of the infant Christ, the *'Menino Deus'* as they called it, or the child-god, which had a long ribbon attached to its waist was placed on the altar. Bates was to say later that an image of this occasion was easily conjured up in his mind whenever he thought of Caripi.

The following day two blue-eyed, red-haired boys arrived and spoke to Bates in English, and presently their father made an appearance. They were a German family named Petzell, living Indian fashion in the forest nearby, about a mile from Caripi. Petzell explained how they came to be there. He had arrived in Brazil thirteen years previously to serve in the Brazilian army then went to America, married in Illinois and settled as a farmer near St Louis. He remained there for a number of years raising a family of five children but, unable to forget the free river-life and perpetual summer of the banks of the Amazons, persuaded his wife to consent to break up their home in St Louis and migrate to Belém.

The joy of this meeting for Bates was that Petzell and his family were expert insect collectors and helped by not only showing him where to find the best specimens but also by collecting many choice examples for him.

> "I rose with the dawn...took a cup of coffee, and then sallied forth after birds. At ten I breakfasted, and devoted the hours from ten until three to entomology. The evening was occupied in preserving and storing my captures. Petzell and I sometimes undertook long

---

remained a slave. It was customary, however, in Roman Catholic lands for the master to manumit (free) both his child and, occasionally, the mother as well. In Brazil, a number of free blacks reached positions of relatively meaningful authority.

Emancipation came to Brazil in a number of stages. Direct Portuguese rule ended in 1822. In 1826, Brazil gave Britain the right to search ships suspected of carrying slaves, and in 1830, Brazil declared the trade to be piracy. Little changed, however, until 1851, when in a well coordinated political and naval campaign, Britain made a conclusive move that cut imports to a mere 3000. In 1871, Brazil passed a partial abolition act; complete emancipation followed in 1888, when the remaining 700,000 slaves were freed.

> excursions, occupying the whole day. Our neighbours used to bring me all the quadrupeds, birds, reptiles, and shells they met with, and so altogether I was enabled to acquire a good collection of the productions of the district."

Vampire bats *(Dysopes perotis,* in the contemporary nomenclature*)* were his next surprise. The room where he slept had not been in use for some time and the rafters were crammed with bats that came out at night in such numbers that they extinguished the light in the room. A vampire bat, swooping silently over its slumbering victim, would administer a painless bite, gorging the blood until it was sated or shooed away. The bleeding was hard to stop, presenting a ghastly sight next morning when wounds were discovered thickly smeared and blotched with blood. Some bats got into Bates's hammock and he found a wound on his hip the next morning, evidently caused by one of them.

After Christmas the orange trees, now in full blossom, attracted large numbers of hummingbirds. Every day, in the cooler early hours of the morning and again later in the evening, they were to be seen whirring about the trees in large numbers. They were very collectable as they were used by taxidermists in England for popular Victorian displays under glass domes. Bates was mesmerised by them:

> "They dart back and forth so swiftly that the eye can scarcely follow them, and when they stop before a flower, it is only for a few moments. They poise themselves in an unsteady manner, their wings moving with inconceivable rapidity, probe the flower, and then shoot off to another part of the tree."

As he walked through the pathways of the forest a hummingbird frequently crossed his path, often stopping suddenly and remaining poised in midair, a few feet from Bates's face. Notes from my own diary from the same area but 160 years later are interesting:

> "Whilst in the forest a large hummingbird was attracted to my red neckerchief - all I heard was the drone of its wings – and a fleeting glimpse of it darting away – the forest is heavy with dew caused by

the high humidity and the constant drip from the canopy far above our heads."[88]

Bates also found plenty of snakes at Caripi, particularly in the branches of trees overhanging rivers. They are most easily seen at night by drifting along the riverbank in a canoe using a light to illuminate their eyes. Animated lianas turn out to be pale-green snakes from time to time, so the rule of the forest is never to catch hold of the undergrowth without looking carefully first.

Bates's excursions with Petzell took them to Murucupi, a creek which passed through the forest, about four miles behind Caripi. Indians and half-breeds who had lived there for many generations in perfect seclusion from the rest of the world inhabit the banks of the Murucupi. On inspection, Bates discovered the native huts were full of children and a number of good-looking mameluco women, who were busily employed washing, spinning, and making farinha. Bates had no difficulty making friends with them as they knew Petzell.

The visitors were invited to share their food and after dinner Bates asked to see more of the creek. A lively and polite old man volunteered to act as guide and they embarked in a small montaria, paddling three or four miles up and down the stream. The creek was about a hundred yards wide, though narrower in some places, and lofty walls of green drapery, with occasional breaks through which were glimpses of palm-thatched huts, masked both banks. The projecting boughs of mighty trees stretched half-way across the creek in some places. They were hung with natural garlands of flowers, and an endless variety of creeping plants clothed the water frontage. Some of these, especially the Bignonias, were ornamented with large gaily coloured flowers. Bates commented that art could not have assorted together beautiful vegetable forms as harmoniously as was here done by Nature. Palms, as usual, formed a large proportion of the lower trees. Some of them, however, shot their slim stems up to a height of sixty feet or more and waved their crowns of nodding plumes in the sky.

Bates noted that he had never experienced such complete darkness, as he and Petzell could not see each other even when walking side by side.

Bates now discovered that a diet including meat was as much a necessity of life in this exhausting climate as it was in Europe. His attempt to live solely on vegetables was a failure, and he could not continually eat the salt-

[88] Author's Amazon diary, *Sixty days on the river*, 2003-2005.

fish and farinha that was the Brazilians' staple diet. After many days without meat of any kind, he asked Senhor Raimundo permission to accompany him on one of his hunting trips to shoot a little game for his own use.

Raimundo lived on the Murucupi, one of the creeks in the forest to which Bates would sometimes walk with Petzell on their longer collecting trips. Raimundo, who had a reputation as a cunning and successful hunter, was known to have secret hunting grounds from which he kept his family supplied with meat. The account that Bates gave of the trip to Stevens, shortly after the event, was brief and unembroidered with reflections:

> "One day and night we were out in a little boat, threading noiselessly by moonlight through winding narrow creeks, with trunks of monstrous trees slanting over, and the broad leaves of the arborescent arums in the swamps gleaming in the moonlight: we had five dogs, and, after a laborious day's work, returned with two pacas and a cutia. The paca is obtained by a person entering the forest with dogs and driving it to the edge of the water, when others remaining in the boat shoot it. On returning, we had to pass over a part of the river clear of islands, ten miles wide, when there was a strong breeze and a heavy swell: we were as near as possible to being swamped, the boat being very small and leaky and the sail heavy. I vowed I would never go on excursions in Indian boats again; but I still enjoyed the trip, and got fresh meat for a week into the bargain."

This was the account of a young man to whom the spoils of the hunt and the pleasure of it were sufficient in themselves. To compare it with the long description of the incident he wrote thirteen years after in *The Naturalist on the River Amazons* is to take a measure of the mental journey that Bates travelled during his years on the Amazons. For the latter is the account of a man with mature ideas on science, life and style.

Details that at the time may have seemed less than important become more so as Bates composes his travel narrative which he is aware needs to be a work of literary quality. He describes himself walking through the woods on the day before the trip, with his double-barrelled gun, a supply of ammunition, and a box for the reception of any insects he might capture. He introduces his characters, Raimundo the taciturn and industrious artisan, his talkative wife Dominga, his apprentice Joaquim, and describes their house, commenting on the conservative attitudes that its primitiveness reflects. Appealing though he finds the relaxed life which nature allows on the

Amazons, he cannot help reflecting that with a modicum of enterprise it might be so much improved. He discovers reasons for the poverty he sees, partly in the lack of adaptability that has prevented the Amazonian Indians from ever thoroughly domesticating animals, and partly in the prevalence among them of a kind of communistic mode of regarding property. The Indian and mameluco people have a fixed notion that their neighbours have no right to be better off than themselves. If any of them have no food, canoe, or weapons, they beg or borrow without scruple from those who are better provided, and it is the custom not to refuse a gift or a loan. There is no inducement, therefore, for one family to strive or attempt to raise itself above the others. There are always lazy people who prefer to live at the cost of their generous neighbours.

After these comments on the way of life of Raimundo and his neighbours, Bates returns to his narrative with one of those sharp sketches of incidents at which he was so adept. In it he portrays the arrival of a procession of Indian pilgrims, led by a cripple with pipe and tabor, who have come begging for alms in the guise of their patron saint, São Tomé. Raimundo receives them with the quiet politeness which comes so natural to the Indian in the position of host, and places their rudely painted banner of the saint in his own oratory, lights candles before it, and he spreads out for them a supper of chicken, fish, and rice. An apprentice waits on the guests with water and towel, and then they are invited to sling their hammocks for the night.

It was now 16th January 1849, and the end of the dry season. The atmosphere became misty; heavy clouds collected where a uniform blue sky had previously prevailed, and down came a succession of heavy showers. The first lasted a whole day and night bringing with it a cacophony of forest sounds that increased in intensity with the coming of the rains.

Bates's journey to Caripi was successful; he was pleased at having amassed a large collection of about twelve hundred species of curious and beautiful insects. He left Caripi on the 12th February, to warm *'adios'* from his hosts. He had passed a delightful time there, in spite of the many privations undergone in the way of food. The wet season had now set in; the lowlands and islands would soon become flooded daily at high water, and the difficulty of obtaining fresh provisions had increased. He decided therefore, to spend the next three months at Belém. There was still much that could be done in that area during the intervals of bright and sunny weather. Once the wet season was over he could embark on another expedition into the interior.

The Journey to Óbidos

In Belém Bates now found himself in a small but active group of naturalists. Wallace had returned from the Guamá at the end of June, and was back with his younger brother Herbert, who had reached Belém on 12th July on board the brig *Britannia* from Liverpool. Herbert, the former trunk maker's apprentice, had shown little inclination for that or any other occupation which could be found for him, and Wallace, feeling the need for a companion after his parting with Bates, had hit on the romantic notion that a young man who failed in England might succeed in the tropics. On the same boat came Richard Spruce, (1817-1893) the botanist, and Robert King, who was to be his assistant. Spruce had heard of the success Bates and Wallace were having with insect collecting on the Amazons, and decided to try his hand botanising there.

Not long after returning to Belém, Wallace was laid up by an infection of his hand, the result of an accidental gunshot wound he received when his gun misfired as he drew it, barrel first, from the bottom of a canoe, something only the inexperienced would do. Advised by a doctor to put his arm in a sling, he was unable to do anything for two weeks, not even pin an insect. Once he had recovered, he assisted Bates in packing their specimens for England. Accompanying this shipment was a letter to Stevens describing their adventure up the Tocantins, portions of which Stevens published in the *Annals and Magazine of Natural History*. He prefaced it with an advertisement to collectors proclaiming the two naturalists as enterprising and deserving young men who had set out on an expedition to South America to explore the vast and unexamined regions of the province of Pará. Stevens exhibited some of their collections at the January 1st meeting of the Entomological Society in 1849, declaring that it included many rare and valuable specimens.[89]

The five young naturalists met frequently during the month they were together in Belém, but there was never any question of pooling resources and working together. The Wallace brothers lived in Nazaré and Bates stayed with a Portuguese family. Spruce and King were guests of the

---

[89] This advertisement appeared in the Annals and Magazine of Natural History (1850), 2nd set. 5. "SAMUEL STEVENS, NATURAL HISTORY AGENT, NO. 24 BLOOMSBURY STREET, BEDFORD SQUARE, begs to announce that he has recently received from South America *Two beautiful Consignments of INSECTS of all orders in very fine Condition, collected in the Para, containing numbers of very rare and some new species ... for Sale by Private Contract.*"

Campbell's, staying either at James Campbell's house in Belém or at Archibald Campbell's estate at Caripi. Spruce was a self-taught naturalist like Bates and Wallace but was treated with unusual consideration by the patricians of Belém. Bates doubtless resented the difference for there is little evidence that he felt any cordiality towards Spruce who is mentioned only once in surviving letters and never in his book. Bates, the Wallace brothers, and Spruce all intended to reach the Upper Amazons during the summer of 1849, but their plans were kept separate, and they set out at different times. A recent biographer of Wallace, Amabel Williams-Ellis in a book entitled *Darwin's Moon* has constructed out of conjecture a joint voyage that never took place.[90]

Not only was there no joint voyage, there was also no occasion when all five were together in Santarém, which as we shall see, Bates visited for a single day in 1849 and then much later on his own. The accounts of the various travellers leave no doubt regarding their individual travels. Wallace, in *Travels on the Amazon,* dates his own departure in August 1849, and a letter from Bates to Stevens, dated 30th August 1849 from Belém, confirms this:

> "Mr. Wallace and his brother have been gone to Santarém half-way to the Rio Negro, or rather more - about three weeks."

As for Spruce and King, it is clear from Spruce's own account that they did not leave Belém until approximately two months later:

> "We embarked on the Tres de Junho on the 10th of October (Bates left on 6th September)."[91]

During these travels, Bates and Wallace became well known to traders, officials, and boat crews up and down the river. News of their activities travelled from settlement to settlement, boat crew to boat crew as the people who lived on and by the river acted like a telegraph system.

---

[90] Williams-Ellis, Amabel, *Darwin's Moon: A Biography of Alfred Russel Wallace.* (London: Blackie, 1966) 45-46.

[91] One of the three great mid-nineteenth century European explorers of South America was Richard Spruce, a commercial collector with no formal scientific training, who worked off and on with Wallace and Bates during the early part of his stay in the Amazon basin. William Hooker was one of his supporters among English botanists, and in 1860 Spruce worked for the British government to collect seeds of the tree used in making quinine; important in combating malaria, then a major barrier to British imperialism. After Spruce's return to England, Wallace, Darwin and William Hooker obtained a government pension for him; Wallace later edited and published Spruce's journals. Wallace, A R, Ed *Notes of a Botanist on the Amazon and Andes*, Vols I and II by Richard Spruce.

The delay in Bates's departure was due in part to his friendship with Dr Ângelo Custódio Correia in Cametá. Dr Ângelo's half-brother, Joaŏ da Cunha Correia, was a river captain and trader who owned a forty-ton schooner, the Santa Maria, in which he planned to make a trading journey to the Upper Amazons or Salimões. A note from the amiable doctor ensured Bates a passage on João da Cunha's vessel. A delighted Bates told Stevens:

> "I shall have an opportunity to proceed to the frontiers of Ecuador, the owner is commander and supercargo both, and tells me he shall have great pleasure in stopping at any place for a day or two, if I should wish to explore it."

João da Cunha planned at first to depart at the end of July. By the end of August, Bates had received yet another lesson in the unreliability of Amazonian transport; the vessel was still not ready for departure. 'I have been miserably delayed', he complained to Stevens. The waiting period was largely taken up in preparations for this journey to the interior.

The only way to travel in this region was along the rivers by means of a variety of small sailing vessels as steam propulsion was still some time away. Traders who resided in remote towns and villages and rarely visited Belém themselves usually owned these small boats. They employed Cabos as Captains of the boats, who were mostly Portuguese half-breeds held accountable to the owner for the cargoes they carried. This trade was usually the exchange of goods for produce among the scattered populations in the interior. It was a form of transport in decline when Bates was on the river mainly because of the difficulty Cabos had in obtaining a crew for any particular journey, something Bates and Wallace had already experienced. Travel on the river was a tedious affair particularly when the wind was blowing in the wrong direction. Then the vessel had to be at anchor for long and uncomfortable periods or the crew could progress only laboriously by means of the 'espia', graphically described by Bates.

The mode of travelling was as follows. The montaria, with twenty or thirty fathoms of cable, one end of which was attached to the foremast, was sent ahead with a couple of hands. They secured the other end of the rope to some strong bough or tree trunk; the crew then hauled the vessel up to the point, after which the men in the boat re-embarked the cable, and paddled forwards to repeat the process.

Plate 7.
A Montaria

Bates's westward journey began on 6th September 1849. He intended stopping somewhere on the northern bank of the river in order to compare the species found there with those from the region of Belém. This comparison revealed the relationship between the fauna of Belém and that of the coastal region of Guiana. He travelled independently and took all that he might need with him and could not expect to obtain during the journey: ammunition, chests, store-boxes, a small library of natural history books, and a hundredweight of copper money. He engaged a boy called Luco to act as servant, described as short, fat, and yellow-faced, whom he had already employed at Belém in collecting. The same night the boat weighed anchor, and on the following day they were gliding along the dark-brown waters of the Moju.

An inconvenient delay to general progress was now caused by the boat's Captain, João da Cunha Correia, who, as he was going to be away from Belém for a lengthy period, decided he would visit Cametá and spend a few days there with his friends without any consideration for the passengers, the condition of the cargo, the vessel, or the crew of twelve. Having arrived at Cametá without incident, João da Cunha remained ashore for twelve days of revelry before departing in the dead of night when both tide and wind were favourable.

Bates critically observed the differences among the crew. One, a young illiterate Portuguese, was as mediocre in manners as the Indians, whose attributes he copied. Another was a tawny-white Camita, the cook was a Cafuzo and the rest were Indians. The pilot, a man of incredible stamina, was from Belém. He remained at the helm for the entire voyage except for a break of two or three hours each morning. The method of running the vessel appeared to be shambolic in Bates's eyes; the crew slept when they felt so disposed and there was no system of watches but somehow everything remained in order in a South American way. Among the Indians was one who had been involved in the rebellion when the Indians and their allies captured Santarém in 1835. He was looked up to by the other Indian crewmembers who always referred to him as the Comandante.

They related of him that, when the legal authorities arrived with an armed flotilla to recapture the town (Santarém), he was one of the last to quit, remaining in the little fortress which commands the place to make a show of loading the guns, although the ammunition had given out long ago.

On 23rd September, they passed Entre-as-Ilhas and made their way across to the eastern shore, the starting point for all canoes negotiating the broad mouth of the Tocantins when going west. Here the waters are more like a passage at sea than on an inland waterway. On 25th September, they sailed westwards along the upper portion of the Pará estuary, which extends seventy miles beyond the mouth of the Tocantins. The wind was fresh, and the schooner rolled and pitched like a ship at sea for a distance of about fifteen miles.

They sailed towards the West with the island of Marajó to their North and, towards evening, entered the narrow strait of Breves where the extensive labyrinth of channels that connects the river Pará with the Amazons commences.

On 26th September, they passed the settlement of Breves. This comprised about forty houses mostly occupied by Portuguese traders. Now the river was only about 400 yards at the widest point, and Bates was much taken with the magnificence of the forest that now pressed down on them.

They travelled for three and a half more days to cover the thirty-five mile length of the Jaburu channel and noted that the banks on either side were hard mud with a covering of thick green mould. His crew told him of the strange superstition that a Pajé, or Indian wizard, whom it was necessary to appease, haunted the waters. If the voyager wished to secure a safe return from the *sertão*, as the interior of the country was called, they should deposit some article on the spot. Here the trees were decked out with all manner of things such as a rag, shirts, straw hats, bunches of fruit, and so forth. Bates noted it was only the Portuguese and uneducated Brazilians who deposited anything. The pure Indians gave nothing, and treated the whole affair with disdain, reclaiming anything they thought worthwhile. As they reached a broad channel called Macaco early in the morning on 30th September they found they had entered a broad stretch of water much less gloomy than the Jacuru, passing through a cluster of islands between which they occasionally caught a glimpse of the main Amazon River. They stopped to gather fruits and saw a flock of scarlet and blue macaws (*Macrocercus Macao* in the nomenclature of the time) feeding on the fruits of a Bacaba palm, looking like a cluster of flaunting banners beneath its dark-green crown.

Bates landed about fifty yards from the tree, and crept cautiously through the forest, but before he reached the birds, they flew making loud harsh screams. At a wild fruit tree he was more successful, as his companion shot an anacã (then classified as *Derotypus coronatus*), one of the most beautiful of the parrot family. It is green, and has a hood of red bordered with blue feathers at the back of its head, which it can elevate or depress at pleasure. The anacã is the only New World parrot that nearly resembles the cockatoo of Australia. It is found in all the lowlands throughout the Amazons region, but is not common anywhere:

> "Few persons succeed in taming it, and I never saw one that had been taught to speak. The natives are very fond of the bird nevertheless, and keep it in their houses for the sake of seeing the irascible creature expand its beautiful frill of feathers, which it readily does when excited."

They collected a quantity of fish, a small alligator and many alligator eggs, which are larger than a hen's and regularly oval, with a rough hard shell. Unfortunately, the alligator was cut up ready for cooking when Bates returned to the schooner, and he could not therefore make a note of its proportions. Each man, being his own cook, skewered the pieces and roasted them over the fire. Bates did not see this particular species of alligator after that.

On 3rd October they reached the main river, Bates rising long before sunrise to view the river by moonlight. He said it had a most magnificent appearance, an imposing vast flowing stream of ochre-coloured turbid waters. By 11 a.m., they arrived at Gurupá, a small village situated on a rocky bank between thirty and forty feet high. This was Bates's opportunity to walk in the neighbouring woods. Intersected by numerous pathways, and carpeted with ground ferns (*Lycopodia*) growing to a height of eight to ten inches, the woods were enlivened by numbers of glossy blue butterflies of the *Theclidae* or hairstreak family. Soon after sunset, they crossed the mouth of the Xingu as a black cloud rose suddenly in the northeast. The Xingu is the first of the great tributaries of the Amazons and is 1200 miles in length. João da Cunha ordered all sails to be taken in as a furious gust of wind burst forth turning the waters into foam, and producing a frightful uproar in the neighbouring forests. Torrential rain followed, but in half an hour all was calm and a full moon appeared in a cloudless sky.

For several days the weather was pleasant, the air was transparently clear with cool and refreshing breezes. On 6th October, they passed a chain of blue hills, the Serra de Almeirim, on the northern horizon. Bates found this sight refreshing after so many miles of flat countryside.

They sailed along the southern shore for several more days, the table-topped hills on the far northern horizon occupying much of their attention. Bates commented on the islands in the river that, being bright green, contrasted with the deep grey mountains in the distance. Soon they came to Monte Alegre, located some ninety miles beyond Almeirim, and built in line with what appeared to be the last summit in the distant mountain range. Here the river bends towards the south as the hilly country recedes to reappear at Óbidos some hundred miles further to the west.

Criss-crossing the river several times to make the most of both wind and tide they continued towards Santarém. The waters became choppy but a gentle wind carried them on to their destination. They reached Santarém on the south bank of the river on 9th October 1849 and, as they drew in to the landing stage, Bates noted the clean and cheerful aspect of the place. He was to spend only one day there on this occasion but met Captain Hislop, an elderly Scotsman who had lived in the area for more than thirty years, and recommended Óbidos as a good place for collecting. Bates, joined Wallace, his younger brother Herbert and Spruce for dinner with the Captain that night, a man who was renowned for giving help to travellers and appears in most travel journals written about journeys undertaken around Santarém between 1840 and 1850.

Bates describes Santarém as having three long streets, with a few short ones crossing them at right angles. There were then about 2500 inhabitants. The town is situated just within the mouth of Tapajós, and is divided into two parts, the town and the aldeia or village. The houses of the whites and traders are substantially built, many with two or three stories, and all whitewashed and tiled:

> "The situation of the town is very beautiful...It is scantily wooded, and towards the interior consists of undulating campos, which are connected with a series of hills extending southward as far as the eye can reach... Although 400 miles from the sea, it is accessible to vessels of heavy tonnage coming straight from the Atlantic...We ourselves had accomplished 200 miles, or about half the distance from the sea, in an ill-rigged vessel, in three days and a half."

On 27th October, Richard Spruce disembarked from one of Captain Hislop's vessels and occupied a house near the Wallaces, though he initially kept his distance. The Amazon buzzed with the activity of English naturalists, but they all tended to stay separately and keep apart unless they were being entertained to someone's house, like the Captain's. Spruce had met Wallace and his brother earlier in Belém, having sailed to Brazil with Herbert from Liverpool, but Wallace was wary of other travelling naturalists following his estrangement from Bates. Spruce said however:

> "I forgot to mention that we have several times seen Mr. Wallace senior when we arrived...he had made up his mind to go to the Rio Negro, but hearing me talk (of) Monte Alegre put it into his head to go thither and he is preparing to start in a few days."[92]

Spruce was surprised that Wallace was shunning other European company but noted that Wallace's Anglophobia had softened as the effect of Captain Hislop's wine soaked evening gatherings elevated everyone's spirits and broke down barriers. On one such evening, Wallace and Spruce struck up their lifelong friendship.[93]

On leaving Santarém, Bates moved further west and on to Óbidos. He had decided this was the best place to stay for a few weeks collecting on the northern bank of the river. Next morning he landed having said goodbye to João da Cunha whom he referred to thereafter as his kind and good friend. Cunha landed the baggage, lifted anchor and continued up the river. Bates was to remain at Óbidos from 11th October until 19th November, the only

---

[92] Richard Spruce to William Hooker, 3rd August 1849, (Letters from Spruce 1842-1890), no. 259, Archives of the Royal Botanic Gardens, Kew.

[93] Slotten, Ross A, *The Heretic in Darwin's Court*, (New York: Columbia University Press, 2004) 59-61.

part of the dry season of 1849 when he was able to devote his time to serious collecting and the study of entomology. Later he spent three weeks there in 1859, but found the place much changed through the influx of Portuguese immigrants and the building of a fortress on the top of the bluff. Bates considered Óbidos one of the most agreeable towns on the great river.

It stood on a layered cliff of *tabatinga* clay, which occurs frequently along the Amazons, hard pink strata alternating with soft yellow, peppered at times with marine shells, and all resting on a solid bed of sandstone. There were about half as many inhabitants as in Santarém, few of whom were pure Indians. Most of the inhabitants were old-established white families, usually with traces of Negro or Indian blood, who owned cacao plantations and cattle estates in the locality. From lack of enterprise, most of them were as poor in cash as they were rich in acres. They lived off the land by a rather simple kind of subsistence farming. The few who showed a little industry had become rich, and Bates observed that Óbidos was a favourite destination of enterprising young men from Belém who came to seek their fortunes by marrying cattle heiresses.

Bates found a room in a town house belonging to one of the most prosperous planters. Its proprietor, Major Martinho da Fonseca Seixas, came from his estate to meet with the English butterfly hunter. He crossed the river in a small boat soon after sunrise, with four dark-skinned paddlers who, at the behest of their master, made the morning air ring with a wild chorus. Bates liked this tall, wiry, and sharp-featured old gentleman of the old school of Brazilian planters. He landed in dressing gown and slippers, and came up the beach chattering, scolding, and gesticulating. Several friends were with him, and after taking coffee, he dressed and went to mass, whilst Bates slipped off to the woods. When Bates returned he found the Major with his retinue seated in hammocks, two by two, slung in the four corners of the room, all engaged in a lively discussion on political questions. They had a demijohn of cashaça and were drinking freely out of little teacups. One of the company was a dark-skinned Cametaense, named Senhor Calixto Pantoja, who was as talkative as the Major. He was a Liberal, while the old gentleman was a rabid Tory. Pantoja rather nettled the old man by saying that the Cametá people had held their town against the rebels in 1835, whilst the whites of Óbidos abandoned theirs to be pillaged by them. The Major then launched out into a denunciation of the Cametaenses and the Liberals in general. He said he was a pure white, a *Mas agonista*; the blood of the Fidalquia of Portugal flowed in his veins, whilst the people of Cametá were a mixed breed of whites and Indians.

Bates's sympathies were probably with the liberals, but he was young and a stranger, and in such discussions he prudently adopted the role of a neutral listener. The outcome was the Major took a liking to him and invited him to his estate, where Bates stayed for two days, housed in a special room that was reserved for the Bishop of Belém when his pastoral duties brought him to Óbidos. The Major was one of the more enterprising estate owners in the region, and possessed the only sugar-mill Bates was to see. However, he found it crude and worked by bullocks, and the old widower's establishment was disorderly and cheerless, managed by a negligent mulatto woman and swarming with dirty Negro children who ran wild everywhere.

The houses all had tiled roofs, and were mostly well built. Bates described the inhabitants as naive but always kind and sociable. Scarcely any palm-thatched huts were to be seen as few Indians lived there. Earlier it had been a Portuguese settlement, and now the better class of population consisted of old established white families, some of whom exhibited traces of Indian or Negro blood or both. During the previous eighty years both Óbidos and Santarém had been places to which considerable numbers of Negro slaves had been imported. Before that time, Indians had been used for forced labour, but their numbers had gradually dwindled away, and they now were an insignificant part in the population.

The chief occupation of the town's population was caring for their cocoa plantations that were located in the low-lying land bordering Óbidos. There was also some attempt at cattle ranching but generally carried out so unenthusiastically the lethargic proprietors earned little money. Bates though was in high spirits and enjoyed the few weeks he spent there. He mingled with the town's inhabitants during the evenings; finding their society more European than Brazilian. The different families met at one another's houses for social amusement. Bachelor friends were not excluded and the whole company, married and single, joined in simple games. Mosquitoes were troublesome at this time of year and so the locals held these gatherings indoors, in sitting rooms rather than out of doors on open verandas.

Sunday was strictly observed at Óbidos...all the shops were closed, and almost the whole population went to church. The Priest, Padre Raimundo do Sanchez Brito, was an excellent old man, and I fancy the friendly manners of the people, and the general purity of morals at Óbidos, were owing in great part to the good example he set to his parishioners.

There was now much to see, the forest teemed with monkeys and Bates rarely saw fewer than four different species daily. The neighbourhood was also rich in insects. Wandering in the broad pathways through the neighbouring forest he came across a magnificent butterfly of the genus *Morpho* with a wingspan of six to eight inches. Called *Morpho hecuba*, it was to be seen daily gliding along at a height of twenty feet or more from the ground. The blue trembled on the wings of this huge butterfly like electricity as it flew about the clearings. Sometimes it seemed to vanish, being hidden by its folded wings which on the underside were shades of grey and brown and impossible to see in the dark undergrowth. Amongst the lower trees and bushes numerous kinds of *Catagrammas*, a group of butterflies peculiar to tropical America, were plentiful.

Bates was pleased with the great variety of beautiful and curious insects found in the woods around Óbidos. For three days a scorching sun had been shining, the river had fallen, and a sandy shoal had emerged from the water in front of his hut. At noon during the hottest part of the day, the whole shoal, a hundred yards long and fifty wide, was covered with butterflies, each one close to the next. Most of them were yellow in colour, belonging to the various species of the big *Papilionidae* and *Pieridae*. Certain large varieties of the *Catopsilia* made a brilliant dash of orange and looked like resplendent ripe fruit in the distance. The butterflies were sucking nutrients out of the fluids in the damp sand, a phenomenon common in South America. If Bates made his way across the shoal and disturbed that tipsy company a yellow cloud would rise before him. The drying puddles on the sand had lured some hundred butterflies of the *Callicora* family. They sunned themselves on the ground and drank the swampy humidity. They were not large, but of a proud beauty, with a metallic green or blue lustre. On the underside of their wings, they all had a design more or less in the form of two figure eights. For this reason, the indigenous people call them *'Eighty-eights'*. Every time they were disturbed they took off and encircled Bates's head in a mobile, gleaming halo, only to resettle on the damp sand in exactly the same place from which they rose. Of this Bates says:

> "They assembled in densely packed masses, sometimes two or three yards in circumference, their wings all held in an upright position, so that the beach looked as though variegated with beds of crocuses. ...All the individuals which resort to the margins of sandy beaches are of the male sex. The females are much rarer, and are seen only on the borders of the forest, wandering from tree to tree, and depositing their eggs on low mimosas, which grow in the shade."

The *Heliconiidae* were to remain Bates's favourites with their fine colouring and long narrow wings. The butterflies of this rather small family are confined to the tropics of the New World, forming an attractive and characteristic faunal element. They are without exception brightly, sometimes startlingly, coloured. In wing shape and markings, they are only likely to be confused with the *Ithomiidae*, another family peculiar to the neotropics. Amongst the morphological characters that separate the *Helicoiidae* from the *Nymphalidae*, with which they have generally been associated as a subfamily, are the following: the long narrow forewing, the wide head, and the simple unbranched humeral (precostal) vein on the hindwing.

The caterpillars, all of simple cylindrical shape and conspicuously coloured, feed on passion flowers. Of the numerous branching spines the caterpillars bear, one pair is on the head, one pair is placed laterally on each thoracic segment, and a further three spines are placed laterally on all the other segments. The chrysalis is greatly contorted, often armed with spines, and generally speckled with gold or silver.

The prevailing ground colour for the imago, the perfect insect, is a deep black, and on this are depicted spots and streaks of crimson, white, and bright yellow or orange, in different patterns according to the species. Their elegant shape, showy colours, and slow, sailing mode of flight, make them attractive objects, and their numbers are so great that they form a positive feature in the physiognomy of the forest, compensating for the scarcity of flowers.

Bates had first encountered the *Heliconiidae* on his arrival in Belém; he was to see them everywhere along the Amazons and its tributaries. He found the genus included a number of clearly defined species, with easily distinguishable colour-patterns, but there were also clusters that included varieties whose position was ambiguous, involving subtle differences. In at least one case he was able to establish that in a stretch of country between two different locations where apparently quite distinct species flourished, a whole series of transitional variations existed; a chain uniting two forms which at first sight seemed only distantly related.

In Óbidos in 1849, Bates first began to observe evolution as a living fact, and at the end of the chapter in *The Naturalist on the River Amazons,* he offered the hypothetical results of his study of the captivating and impressive *Heliconii*. Later in 1862 he was to write:

"...since the publication of the Darwinian theory of the origin of species, it has been rightly said that no proof at present existed of the production of a physiological species, - that is, a form which will not interbreed with the one from which it was derived, although given ample opportunities of doing so, and does not exhibit signs of reverting to its parent form when placed under the same conditions with it. Morphological species, - that is, forms which differ by an amount that would justify their being considered good species, have been produced in plenty through selection by man out of variations arising under domestication or cultivation. The facts just given are, therefore, of some scientific importance; for they tend to show that a physiological species can be and is produced in nature out of the varieties of a pre-existing closely allied one. This is not an isolated case; for I observed, in the course of my travels, a number of similar instances. But in the very few has it happened that the species which clearly appears to be the parent coexists with one that has been evidently derived from it."[94]

When Bates's Linnean paper was published in 1862, Darwin seized on these ideas because he could at last see factual evidence supporting his theory of changing patterns of existence within a period that could be easily understood by man. '*I hope*' he said, '*in my future work to profit by them and make use of them*'. The references to mimicry in insects in *The Descent of Man* are particularly interesting, since they examine a topic which became a key ingredient in Darwin's thinking on evolution and natural selection at that time. Mimicry in insects was to Darwin supporting evidence for his theories, an opinion confirmed by his statement in *The Descent of Man*:

"Since the publication of Mr. Bates'(s) paper, similar and equally striking facts (about mimicry) have been observed by Mr. Wallace in the Malayan region, by Mr. Trimen in South Africa, and by Mr. Riley in the United States."

As Janet Browne comments in *The Power of Place*, mimicry in butterflies was 'destined to become evolution's most elegant practical support'.[95]

[94] Slotten, Ross A, *The Heretic in Darwin's Court*, (New York: Columbia University Press, 2004) 59-61.

[95]Darwin Correspondence Project: Extract from Letter 3100: Darwin, C R to Bates, H W, 26th March (1861) Down Bromley Kent. "I have read your papers with extreme interest & I have carefully read every word of them. They seem

It now started to rain with showers that presaged the great rains of November and Bates decided that he must continue on his next leg upriver before the waters became too high for comfortable travel. He met a trader named Penna who was about to leave for the Rio Negro, and Bates arranged for a passage with him, Penna agreeing to give him part of the fore-cabin of his boat. Bates bought new provisions, moved his equipment on board and they set sail on 19th November 1849.

Penna was a general trader who would set off up the river system in his boat laden with goods of every description that the local people on the river and in the settlements he passed might need. He would stay afloat as long as necessary to trade all this merchandise, stopping at individual dwellings along the river. All his trade goods were exchanged for the natural products of the forests and the river such as cocoa and rubber. He would then set off downstream to sell his goods to the exporting warehouses in Belém. Time was therefore of little or no importance to Penna and he tended to loiter along the river, to stop and stay if trading was good. His wife and two children were with him so that he avoided the possible temptations of lonely travel. Caterina, his wife, was a Mameluco woman of pleasant temperament. At first Bates was pleased that this was to be a slow voyage as it would afford him the opportunity to collect in many more places than normal journeying would allow. However it turned out to be too slow and he eventually became impatient with the Pennas as the rains got worse and the journey to Manaus took more than two months. When he wrote to Stevens at this time, he complained at the miserable slowness of everything, but later when he wrote up his travels in *The Naturalist on the River Amazons* he gave the impression that he had had a delightful time. The truth of the matter is probably that as all tropical travellers experience times of boredom and exhilaration, so did Bates. It was the highlights he emphasised to add colour to his narrative.

In fact, the boat did not have sufficient crew for the journey they were undertaking. There was a mullato, a cafuzo and an Indian plus Bates's boy Luco who had to assist in rowing. When there was a good wind behind it the boat made progress but otherwise it made hard work of its passage up a river which was some ten feet above its normal level, with exceptionally strong down currents caused by the rains. The strong current forced the boat to cross and recross the river to take advantage of the winds but often it slipped back downriver by as much as a mile at a time. Penna was

---

to me to be far richer in facts on variation, & especially on the distribution of varieties & subspecies, than anything which I have read. Hereafter I shall reread them, & hope in my future work to profit by them & make use of them."

unconcerned as he was happy to trade at the slowest possible pace and to go ashore every day for the mid-day meal. So the long and slow ascent to Manaus continued.

The Indian, Manoel, would rise at the first glimpse of sunlight to show above the long dark line of forest, his clothes and hammock soaked with dew as he slept outdoors rather than in the stifling cabins. Manoel would plunge into the river to revive himself as was the habit of Indians of both sexes, sometimes for warmth's sake as the temperature of the water was often considerably higher than that of the air.

> "Penna and I lolled in our hammocks, while Katita [Bates's name for Caterina] prepared the indispensable cup of strong coffee, which she did with wonderful celerity, smoking meanwhile her early morning pipe of tobacco."

Then depending on the wind, they made some movement but if there was none they could only progress by espia. Despite the slow progress Bates commented:

> "We generally ceased travelling about nine o'clock fixing upon a safe spot wherein to secure the vessel for the night. The cool evening hours were delicious; flocks of whistling ducks (*Anas autumnalis*), parrots, and hoarsely screaming macaws, pair by pair, flew over from their feeding to their resting places, as the glowing sun plunged abruptly beneath the horizon. The brief evening chorus of animals then began, the chief performers being the howling monkeys, whose frightful unearthly roar deepened the feeling of solitude which crept up as darkness closed around us. Soon after, the fireflies in great diversity of species came forth and flitted about the trees. As night advanced, all became silent in the forest, save the occasional hooting of tree frogs, or the monotonous chirping of wood-crickets and grasshoppers."

For three days they made little headway, but on the night of 22nd November the moon appeared in a misty halo suggesting a balmy night ahead. Suddenly however at around eleven o'clock a great storm arose which almost sank the coberta, hurling it at the shore. Dark clouds gathered over the scene and lightning flashed. To ensure the vessel was not dashed to pieces they needed all hands. Bates had not experienced anything as startling as this before:

> "...a few yards farther on, where the shore was perpendicular and formed of crumbly earth, large portions of loose soil, with all their super incumbent mass of forest, were being washed away; the uproar thus occasioned adding to the horrors of the storm."

After about an hour the storm abated but the deluge of rain continued until early in the morning and almost non-stop flashes of white lightning continued throughout the night. Next morning everything was comprehensively soaked and a grey gloom of clouds hung over the river and forest.

On 26th November, they arrived at a large sand bank linked with an island in mid-river, in front of an inlet called Maracá-uaçu. They went ashore for what turned out to be an enjoyable stopover, the island being much more open than the depressingly close forest. Penna had stopped simply to play with the children on the sands and to allow Caterina to wash their clothes.

Bates weaves a colourful story for the reader from these scenes:

> "In wandering about, many features reminded me of the seashore. Flocks of white gulls were flying overhead, uttering their well-known cry, and sandpipers coursed along the edge of the water. Here and there lonely wading birds were stalking about; one of these, the Curicaca (*Ibis melanopis*), flew up with a low cackling noise, and was soon joined by a unicorn bird (*Palamedea cornuta*), which I startled up from amidst the bushes, whose harsh screams, resembling the bray of a jackass, but shriller, disturbed unpleasantly the solitude of the place. Amongst the willow bushes were flocks of a handsome bird belonging to the *Icteridae* or *troupial* family, adorned with a rich plumage of black and saffron-yellow. I spent some time watching an assemblage of a species of bird called by the natives Tangurupará on the Cecropia trees. It is the *Monasa nigriforns* of ornithologists, and has a plain slate-coloured plumage with the beak of an orange hue. It belongs to the family of Barbets, most of whose members are remarkable for their dull, inactive temperament. Those species, which are arranged by ornithologists under the genus *Bucco*, are called by the Indians, in the Tupi language, *Tai-assu uira*, or pig-birds. They remain

> seated sometimes for hours together on low branches in the shade, and are stimulated to exertion only when attracted by passing insects. This flock of Tamburi-para were the reverse of dull; they were gambolling and chasing each other amongst the branches. As they sported about, each emitted a few short tuneful notes, which altogether produced a ringing, musical chorus that quite surprised me."

On 27th November, they crossed the provincial boundary between Pará and Amazonas near the wooded promontory called Parintins. Here a small canoe weighed down with tobacco, which the owner, a freed Negro named Lima, was taking downriver to Belém to barter for European goods, passed them. These would eventually be taken upriver once more, as was the method of trading in these parts. Bates, always observant, notes that the Negro was from Pernambuco but belonged to the Mundurucu nation. A little girl of the Maués tribe accompanied him. This tribe is considered by ethnologists to be a remote branch of the Mundurucu, but which, in a manner that seems to have been general with the Brazilian aborigines, had adopted different customs and a different language due to long isolation. Penna told Bates there were scarcely two words alike in the languages of these two peoples, although there were words closely allied to Tupi in both.

We now see Bates taking an intense interest in the people he meets, something he would continue to do throughout his voyaging. He thought the little girl had something of the savage about her appearance:

> "Her features were finely shaped, the cheekbones not at all prominent, the lips thin, and the expression of her countenance frank and smiling. She had been brought only a few weeks previously from a remote settlement of her tribe on the banks of the Abacaxi, and did not yet know five words of Portuguese. The Indians, as a general rule, are very manageable when they are young, but it is a general complaint that when they reach the age of puberty they become restless and discontented. The rooted impatience of all restraint then shows itself, and the kindest treatment will not prevent them running away from their masters; they do not return to the malocas of their tribes, but join parties who go out to collect the produce of the forests and rivers, and lead a wandering semi-savage kind of life."

They spent the night at Serra dos Parintins and as they rose the next morning Bates went into the forest to shoot monkeys, though the sortie was unsuccessful. He noted how the virgin forest was extremely varied with great tracts of lianas coiling their way from tree to tree hindering movement. All around him he found the most beautiful ferns and lichens, so much so that he described the place as a museum of cryptogamic plants. He also discovered some longicorn beetles and a large kind of grasshopper whose wings resembled a leaf providing the insect with a complete disguise. Both these insects were of great interest to him especially as the latter was an example of cryptic colouration.

Bates arrived the next evening at the small settlement of "Vil Nova" (now Parintins), where he stayed for the next four days. Each day he went walking along the shore to collect insects from the great variety of lovely little butterflies found on the thick carpet of flowering shrubs. It was now 3rd December, and he saw many fascinating birds, in particular waders and hawks.

On 6th December, two days after leaving Villa Nova, Penna stopped at a small settlement lying below a high cliff of *tabatinga* clay called the Barreiros de Carauaçu, on the island of Tupinabarana. A festival celebrating the feast of Our Lady of Conception was in progress, and the general population took advantage of the situation created by Penna's arrival to trade their produce, – salt fish, oil of manatee, and fresh fruit – for articles that would make their fiesta more enjoyable. There were about sixty people present, mostly civilized Tapuyo Indians and a few mamelucos. Deciding that Penna had a better Madonna than theirs, they arranged to borrow it, taking it ashore in a parade to the sound of salutes made by their crude hunting rifles. Penna and Bates were invited to take part in the fiesta, which, although it lacked the riches and spectacle of those Bates had seen in Belém, had a great deal of rural charm. No priest was present, and the Indians satisfied themselves with a litany and a hymn. Afterwards they sat down on large mats to a supper consisting of a giant boiled pirarucu, five feet long, stewed and roasted turtle, farinha, and bananas. They began to drink a spirit distilled from mandioca cakes and, when they were adequately relaxed, to dance. Bates observed the dances were European rather than Indian, varieties of the *lundu*, an erotic dance similar to the fandango, originally learnt from the Portuguese.

He found the festivities contained that strange mix of the European and the indigenous which is so often encountered in South America, something that has its origins with the Jesuits as they proselytised the indigenous people of Brazil. The festival was exotic and ended with much dancing and drinking, the whites and mamelucos holding their festivities at one end of the village, the Negroes drumming their African rhythms at the other. The young Indian women went to the White's ball, the elder Indian women went to watch the younger ones dance, but the Indian men attended the Negroes' dance. Bates wandered between the two parties watchful and curious but he says nothing about how far he succumbed to the revelry.

During the ten days that Penna delayed at the Barreiros de Carauaçu, Bates added substantially to his collections. On 16th December, they continued their voyage along the northern shore. Six days later they passed the mouth of the most easterly of the numerous channels which lead to the large interior lake of Saraca, and on 23rd December 1849 threaded their way through a series of passages between islands when they again saw human habitation. This was ninety miles distant from the last houses they had seen at Carauaçu. One day later they arrived at Serpa. It is interesting to note that Wallace had been here the day before, hurrying on his way up the river towards Manaus. He had in fact arrived at Vila Nova after Bates and, travelling in the dark, must have passed him unnoticed on the river.

He seems to have found his stay at Serpa pleasant and made friendships that were to last for the remainder of his stay in Brazil. One such acquaintance was a planter named João Trindade who was an old friend of Penna. When Bates prepared to leave Serpa after the Christmas celebrations, João Trindade invited the whole party to spend a few days on his estate. They set out once more on 29th December passing through further archipelagos until, on New Years Eve, they emerged into open waters. Here they saw to the south the sea-like expanse of water where the Madeira, the greatest of all tributaries, united its waters with those of the even greater Amazon. Bates stared in total awe at the size of the rivers; it was as though he had entered the high seas:

> "I was hardly prepared for a junction of waters on so vast a scale as this, now nearly 900 miles from the sea. Whilst travelling week after week along the somewhat monotonous stream, often hemmed in between islands, and becoming thoroughly familiar with it, my sense of the magnitude of this vast river system had become gradually deadened; but this noble sight renewed the first feelings of wonder."

They were on route to the sítio belonging to João Trindade that stood on a tract of land just above the high water level of the Amazons. It was located on the most fertile land in the region, mingling rich alluvium with thick vegetable mould. João had settled here long ago but the Mura Indians destroyed his original farm during the rebellion of 1835.

What Bates now saw was the result of a decade of reconstruction and he was impressed, regarding his host as a quite extraordinary exemplar of inventiveness in a region where most men lived for the survival they could gain by the smallest amount of work they could undertake. João Trindade, in fact, satisfied both Bates's apparently paradoxical ideals for he seemed to have combined the endeavor of a good Leicester merchant with the freedom of life and manner of the forest dweller that Bates found so attractive. He was not merely a planter but also trader, fisherman, and boat-builder. When Bates arrived, a large sailing canoe stood in the stocks, sheltered by a large shed. All this João Trindade accomplished without employing slaves; his farm was worked by the free labour of his family, his godsons, an emancipated Negro and a few Indians who worked around the estate and fished for the family's needs. The year was now 1850.

This was to be remembered by Bates as a golden moment in his travels on the Amazons. João Trindade not only extended to him the genuine hand of friendship, but was extremely hospitable. He was also one of the few people on the Amazons Bates met who had a mind flexible enough to comprehend the real purposes of Bates's research. He sent his godsons to accompany Bates on his excursions in the forest to show him places where monkeys abounded and butterflies were to be found in profusion.

They left on 8th January 1850, and the following afternoon arrived at Amatari, a miserable little settlement of Mura Indians. Here the indigenous people were anything but friendly. They lived in hovels and made no effort to cultivate the land around them. Inside one of the hovels women were cooking a meal of fish, the stinking entrails of which were scattered on the floor of the living area. The bodies of the naked children were smeared with a black mud that they said was to protect them from the ravages of the abundant mosquitoes. Young men were lounging about seemingly with little purpose, though Bates was amazed at their broad chests and wonderfully thick and muscular arms. The place was dirty and the foul smell from the fish entrails was emphasised by the hot and humid atmosphere. Bates was glad to leave for the people had offered no civilities and he found their appearance savage and menacing. Penna had declined to trade with them as they had nothing to barter and departed to the strains of hoots and shouts of anger from all on the shore.

It was at this point that Bates found he had been robbed, he did not know exactly where it had happened but he thought somewhere between Óbidos and Amatari, possibly whilst he was celebrating Christmas at Serpa or at Amatari. He had been relieved of most of the Brazilian money he carried which was to have paid for his expenses until he returned to Belém. There was now no way he could obtain funds from Singlehurst & Co in Pará (Belém) within any reasonable time, and, shy of asking for credit he was left with only £11 for the expenses he would incur at Manaus. After all the toil of the arduous journey he had just completed he was disappointed to realise that he would not be able to stay there until the weather improved but would instead be obliged to return to Belém without making a collection.

They now continued their journey along the northern shore and after six days sailing, on the 14th, passed the upper mouth of the Paraná-mirim da Eva, which was an arm of the river created by an island some ten miles long lying parallel to the shore. From here it was a mere 20 miles to the Rio Negro. The weather was wretched until the 18th and Penna was eager to complete his trading by the time they arrived at Manaus. Here Bates discovered a new pest to torment him for the remainder of the journey, the Pium fly (black-fly) of which he says:

> "I was aware of the presence of flies; I felt a slight itching on my neck, wrist, and ankles, and, on looking for the cause, saw a number of tiny objects having a disgusting resemblance to lice, adhering to the skin. This was my introduction to the much-talked-of Pium. On close examination, they are seen to be minute two-winged insects, with dark coloured body and pale legs and wings, the latter closed lengthwise over the back. They alight imperceptibly, and squatting close, fall at once to work; stretching forward their long front legs, which are in constant motion and seem to act as feelers, and then applying their short, broad snouts to the skin. Their abdomens soon become distended and red with blood, and then, their thirst satisfied, they slowly move off, sometimes so stupefied with their potations that they can scarcely fly. No pain is felt while they are at work, but they each leave a small circular raised spot on the skin and a disagreeable irritation."

Manaus was built on a tract of elevated, but uneven land, on the left bank of the Rio Negro. In 1850 it had about 3000 inhabitants but had previously been a fortification used by the Portuguese whilst they were collecting slaves off the Rio Negro. The most warlike of the tribes from which the slaves were collected was called the Manaus, hence the modern name for the place. The Portuguese captured hundreds of these slaves and when Bates arrived in 1850 many of the inhabitants could still remember these events. However South American Indians were particularly susceptible to European diseases and as slaves they were replaced by the much hardier Africans. Slavery was to continue in Brazil until 1888.

Bates liked Manaus and found the climate to his liking. There were few insect pests in the town to bother him:

> "...the soil is fertile and capable of growing all kinds of tropical produce (the coffee of the Rio Negro, especially, being of very superior quality), and it is near the fork of two great navigable rivers. The imagination becomes excited when one reflects on the possible future of this place, situated near the centre of the equatorial part of South America, in the midst of a region almost as large as Europe, every inch of whose soil is of the most exuberant fertility, and having water communication on one side with the Atlantic, and on the other with the Spanish republics of Venezuela, New Granada, Ecuador, Peru, and Bolivia."

Bates considered the town had seen better times as now it was in a most wretched plight, suffering from a chronic scarcity of the most necessary and basic food. The attention of the settlers was formerly devoted almost entirely to collecting the wealth of the forests and rivers so agriculture was neglected. The neighbourhood was now not even producing sufficient mandioca-meal for its own consumption. Much of its food and luxuries was imported from Portugal, England, and North America.

Once ashore, he paid his respects to the leading trader of the town, an Italian of long residence known to the Brazilians as Senhor Henrique Anthony. The good-natured merchant, whose kindness to strangers is celebrated in every book of Amazonian travels belonging to this period, solved the first of Bates's problems by offering him a couple of rooms in his own house. Bates now met up with Wallace who had arrived three weeks before and they exchanged notes on their travels. Wallace was exploring with his brother Herbert who had arrived from England in July 1849 but was alas finding the life of a travelling naturalist not to his liking. Richard Spruce, who was to become a well-known botanist, had arrived on the same ship as Herbert and they all appeared to enjoy each other's company for the brief time they were together. The four of them were entertained at Senhor Henrique's table so Bates was relieved of his immediate need for money and it is even possible that Wallace helped his old colleague solve this problem. W H Edwards says of Senhor Henrique in 1847:

> "His house was always open to passing strangers, and others beside ourselves were constantly there: enjoying his hospitality. Both the Senhor and his lady showed us every attention, and seemed particularly anxious that we should see all that was interesting or curious in the vicinity, while they constantly kept some Indian(s) in the woods for our benefit. The Senhora was an exceedingly pretty woman, about twenty-two, and delighted us by her frank intercourse with strangers; always sitting with them at the table, and conversing as a lady would do at home...She had three little girls, Paulina, Pepita, and Lina, with a little boy of four years, Juan. All these children had light hair and fair complexions, and the blue-eyed baby Lina especially was as beautifully fair as though her home had been under northern skies...Each of the children had an attendant; the girls, pretty little Indians of nine or ten years, and Juan, a boy of about the same age. It was the business of these attendants to obey implicitly the orders of their little mistresses and master, and never to leave them. Juan and his boy spent much of their time in the river, taking as naturally to the water as young ducks."

By now the heavy rains had set in. There were sunny intervals, during which Bates and Wallace would ramble together through the beautiful open woods around Manaus, or take the forest road to the waterfall two miles out of the town. The townspeople made picnic excursions to the waterfall and the gentlemen and possibly the ladies spent the sultry hours of midday bathing in the cold and bracing waters. It still exists but is now neglected and polluted. However, in this season, there were no bathers of either sex, and, charming as the woodlands were, they provided a miserably small harvest to the collector. Even when specimens were obtained, it was difficult to preserve them, since the atmosphere became so humid that insects grew mouldy and bird skins lost their feathers. Bates and Wallace reconciled themselves to accepting this as a period of relaxation before starting on further but separate expeditions. Wallace was determined to make a long voyage up the Rio Negro to its upper reaches whilst his brother, who had now finally decided to give up collecting altogether, was to remain a few further months at Manaus, return to Belém and thence to England.

Decaying gently like so many places on the river all this was to change in 1853 when Barra was renamed Manaus and became the capital of the Amazonas region. Life here was anything but boring for Bates and Wallace during their brief stay for the company was international. Hauxwell, the English bird-collector, was there, famous for the flawlessly preserved skins he had sent to half the museums of Europe. Accompanying him was a young Indian assistant who was expert in the use of the blowpipe (*zarabatana*), the principal weapon of many of the indigenous tribes of the Amazon. About nine feet long, the blowpipe was constructed from the stem of the palm *Iriartea setigera,* which varied in diameter from the thickness of a finger to two inches. After these stems were carefully dried, the soft pith was bored out and the bore polished smooth. Its mouthpiece was then fitted and the whole smeared with beeswax. The poison darts that were to be used in the blowpipe were shaped from the hard rind of palm-leaf stalks. Thin strips were cut and made sharp as needles by scraping the ends with a knife or an animal's tooth. These were than dipped in the poison called curare, concocted from the bark of the vine *(Strychnos voxifera).* To aid in its trajectory, each dart was furnished with a little oval tuft of a silky material from the seeds of the silk-cotton tree *(Eriodendron mmauma).* In an inexperienced man's hands, this contraption was unwieldy, but adult Indians in the upper Amazon used the blowpipe with remarkable adeptness and to deadly effect at a bird or other animal.

The Irishman Neill Bradley and the American Marcus Williams were also in Manaus at that time and formed a partnership of wandering traders, among the most enterprising on the river. There were three Germans, one of whom spoke English well and was an amateur naturalist. Oddest of all was the deaf and dumb American called Baker, who had travelled through Peru, Chile, and Brazil, paying his way by selling the deaf and dumb alphabet, with explanations in Spanish and Portuguese. He supplemented his living with phrenology, and gained considerable prestige among Brazilians because whenever he read their heads he would write on his slate, 'Very fond of the ladies', an observation his customers felt to be penetratingly true.

Bates sang with the Germans, perfected his knowledge of their language, and from Julio, the Indian employed by Hauxwell, took lessons in native methods of hunting.

> "I have now got blowpipes and poisoned arrows, (he told Stevens). This is a capital way of killing birds for specimens. Ammunition is cheap; specimens are perfect."

Bates continues:

> "After we had rested some weeks in Manaus, we arranged our plans for further explorations in the interior of the country. Mr. Wallace chose the Rio Negro for his next trip and I agreed to take the Salimões. My colleague has already given to the world an account of his journey on the Rio Negro, and his adventurous ascent of its great tributary, the Uapes. I left Manaus for Ega, the first town of any importance on the Salimões, on the 26th of March, 1850."

Chapter 6

The Journey to Ega 1850

Plate 8.

Latitude 3 deg, 20 min, 56.7 secs South – Longitude 64 deg, 42 min, 48.4 secs West = Tefé

Bates continued his journey westwards from Manaus towards the upper Amazons or Salimões. It would take 35 days to reach Tefé some 400 miles further up the river. He embarked on a coberta that was returning to Tefé, considered at that time to be the only town of note on the Salimões. The owner of the coberta was an old, white-haired Portuguese trader from Tefé called Daniel Cardozo. He was returning home having delivered a shipment of turtle oil to Santarém. At the same time he had been attending the assizes as a juryman, a public duty he performed without payment, which had taken him away from his business for more than six weeks. Bates had so much baggage that Cardozo suggested it should be transported by separate coberta and that Bates should join him in a small boat that was to travel in tandem with the cargo. Cautious as ever, Bates preferred to remain with his equipment, looking forward to the many opportunities this would afford for landing and making collections on the banks of the river. Bates also shipped his collections made between Belém and the Rio Negro, in a large cutter that was about to descend the river for Belém. After a hard day's work, all his chests were stowed on the Tefé-bound coberta by eight o'clock that night. The crew embarked and, despite the drunken condition of some of them, the vessel set off on what Bates described as a clear and still night on the smooth waters of the river.

As they had to proceed by espia along the left bank of the Salimões, the journey became difficult. The middle of the waterway was full of debris and floating hazards caused by the rainy season that had begun in the West. All day long the crew and Bates, taking turns at the laborious task, had to haul the coberta along the shore.

On 28th November they passed the mouth of Ariaú, a narrow inlet which, emerging in front of Manaus, links with the Rio Negro. Here the vessel was nearly sucked into the rising waters of the Ariaú, an event that was only prevented by all on board making the tremendous effort needed for the espia.

Progress was slow and, from time to time, members of the crew supplemented their meagre rations with fresh caught fish. They caught a ten foot long manatee, *Trichechus inunguis,* which caused much celebration. They stopped for several hours, building a fire on the shore and barbequing the animal. Bates said it tasted of coarse pork but the greenish coloured fat was inedible and had an unpleasant fishy flavour.

Confinement to the canoe, the trying weather, frequent and drenching rains with gleams of fiery sunshine and the woeful desolation of the river scenery made this a wearisome journey, but Bates still comments that overall he enjoyed it. There were few mosquitoes to trouble him and sleeping on the deck well wrapped in a sail or blanket made the nights acceptable. Even so he was becoming increasingly exhausted. His narrative hints at how he was feeling: 'In passing slowly along the interminable wooded banks week after week', he begins as he describes the shore and the passing forests.

Despite the dreariness, the sight of the great variety of palms occasionally aroused Bates's interest. On the 22nd April 1850 he arrived at the Paraná-mirim of Arauanaí, a narrow strip where they rowed for half a mile through a bed of glorious Victoria water lilies, *Victoria amazonica*, where flower-buds were just beginning to develop.

After seven more days without incident they arrived on 30th April at a narrow opening on the riverbank that could be mistaken for some insignificant stream. It was in fact the mouth of the Tefé river, where Tefé was located.

> "After having struggled for thirty-five days...I had now reached the end of the third stage of my journey, and was now more than half way across the continent. It was necessary for me, on many accounts, to find a rich locality for Natural History explorations, and settle myself in it for some months or years. Would the neighbourhood of Ega turn out to be suitable, and should I, a solitary stranger on a strange errand, find a welcome amongst its people?"

Plate Section i: [European collectors in a] "Panoramic view of Rio de Janeiro taken from the Aqueduct"

Plate Section ii: "Virgin Forest near to Manqueritipa in the province of Rio de Janeiro"

Plate Section iii: "Rio Panahyba"

Plate Section iv: "Mouth of the River Caxoera"

Plate Section v: "A meeting of Amerindians with European Travellers"

Plate Section vi: [Jaguar] “Hunt in a virgin Forest”

HABITANS PECHEURS

Côte des Jlheos.

Plate Section vii: “Native Fishermen” [with a Cayman]

Plate Section viii: "Feast of Saint Rosalie, Patroness of Negroes" [Slaves]

Early the next morning, on 1st May, the men resumed their rowing and brought the craft to the shore. This passage was through the lake of Tefé, a smooth five-mile wide stretch of the river, clear of islands and curving away to the west and to the south. To their left was a gentle grassy slope and in the centre of a broad river junction was the town of Tefé, no more than a cluster of about a hundred houses, mostly palm-thatched with whitewashed walls, a few with red roof tiles and each with its neatly-enclosed orchard of orange, lemon, banana, and guava trees: Bates comments:

> "We let off rockets and fired salutes, according to custom, in token of our safe arrival, and shortly afterwards went ashore."

Bates's first impressions were good and he contemplated spending some time there. On going ashore, he was immediately made welcome and an ox was killed for a feast in his honour. Bates met the Delegado of police, Senhor Antonio Cardozo who welcomed him cordially. Cardozo was from Belém, having originally arrived at Tefé as a moderately successful trader. The senior army officer at Tefé, a mulatto called Praia, also proved to be friendly. His wife Dona Anna, from Santarém, was the leader of fashion in the settlement. Bates also met the Padre, Father Luiz Gonçalves Gomez, a nearly pureblood Indian and native of one of the neighbouring villages, who had been educated at Maranham, a city on the Atlantic seaboard.

After finishing his rounds of visiting and introductions he went to meet a venerable native merchant, Senhor Romão de Oliveira. A tall, corpulent, fine-looking old man, de Oliveira received Bates courteously. He placed his house and store at Bates's disposal and, for some time afterwards, refused to accept payment for the goods Bates had purchased. Both Tefé and its people compensated for the rough journey which Bates had previously described to Stevens as 'intolerably wearisome' and later in *The Zoologist* as 'very tedious'.

However, Bates was still depressed. His personal servant had robbed him of everything, including his shoes. For nearly a year, he had received no money from the sale of his specimens, and life in such abject poverty had become unbearable. Miraculously, a boat from Belém arrived with £40 for him, and two months later another £30, along with more letters from his father beseeching him to come home.

He was also getting tired of the undertaking. In a letter dated 23rd December 1850 and reprinted in Vol VIII of *The Zoologist,* he wrote to Samuel Stevens:

> "Considering the unsatisfactory nature of my future prospects in this profession, I think I do better in returning to a more certain prospect of establishing myself. I have now, therefore, only one idea, that of returning to England."

In the same letter, he half-heartedly proposed that he might venture to the Amazon sometime again in the future, going to Peru or Venezuela after making better arrangements. At the end of the letter, there is a hint of envy:

> "Mr. Wallace, I suppose, will follow up the profession, and probably will adopt the track I have planned for Peru; he is now in a glorious country, and you must expect great things from him. In perseverance and real knowledge of the subject, he goes ahead of me, and is worthy of all success."

Bates decided to return to Belém. On arrival he learned from Stevens that his collections were selling well and that a sizeable amount of money awaited him. He was also acquiring a coterie of admirers because of the extracts of his letters published by Stevens in *The Zoologist.* One man had written that:

> "These letters appear to me to be distinguished by a devotion to science, and by an enthusiasm, which are seldom exhibited...His descriptions depict the primeval forests of South America, where, in not a few instances, no European foot but his own seems ever to have trod, with a glowing freshness and vivacity which brings everything in the clearest manner before the eye of the mind."[96]

In addition, Stevens reported that a new species of butterfly that Bates had discovered had been named *Callithea batesii.* All this good news helped him to recover his lost composure

In June 1851, he wrote to Edwin Brown: 'I am here now in Belém, working very hard, and thinking what is best to be done next. After all I think I shall not be able to settle at home in a quiet life'. Of course, this was written before he caught yellow fever.

---

[96] Letter to Stevens, 16th March 1849, *Zoologist* 8 (1850) 2664.

Wallace was now in Manaus, stranded like Richard Spruce as they each tried to put together a crew for their journeys up the Rio Negro. Wallace got on well with Spruce who offered to share his accommodation with him for what they both hoped would be a brief period. Wallace received a bundle of mail containing a letter postmarked in early June, from Daniel Miller, now the British consul in Belém, telling him that his brother Herbert was seriously ill with yellow fever and was unlikely to recover. Wallace was shocked by this. He had no idea that Herbert was still in Belém, believing that he had left Brazil in February or March of 1851 and was well on his way home. The next letter he received reported the death of Miller from an attack of 'brain fever' but he learned nothing more about his brother's illness.

Wallace continued with his tasks as though his brother's illness was a minor problem, yet it must have been constantly on his mind. In the end, Wallace opted to remain where he was, complete his preparations for his travels, and not to return to Belém. No letters describing his thoughts or feelings on the subject are preserved and he provides no further clues in his autobiography. What he does say is that he spent these weeks buying supplies, packing his miscellaneous collections and arranging for his expedition to the Uaupés. At night, he immersed himself in the pleasures of rational conversation with Spruce, claiming this to be the greatest and rarest of pleasures.

In part, his choice was limited by economic necessity; he had to continue his explorations to earn a living. Vessels to and from Belém were infrequent, and he would later learn that Herbert's ship had been delayed for many weeks by necessary repairs. Wallace was to remain silent on the matter of Herbert's death and it is not clear exactly when he learned of it.

The details of Herbert's final days come from a surprising source. Bates had arrived in Belém not long after Herbert, having all but abandoned his career as a traveling naturalist. Not long after he returned to his accommodation, he fell ill with fever and began vomiting. He sent a servant to town to buy medicine, while he spent several nervous hours pacing back and forth on the veranda wrapped in a blanket. He drank various teas and then a 'good draught' of a concoction of elder blossoms to induce sweating, before falling insensible into his hammock. He did not wake until midnight, by which time he was weak, with every bone in his body aching. He purged himself with Epsom salts. Forty-eight hours later his fever broke and in another eight hours he had completely recovered. Previously he had nursed the ailing Herbert Wallace on his deathbed. The difficult task of informing

Herbert's mother of her son's death fell to him, since Wallace, the older brother, was out of reach, somewhere in the wilderness.

Bates writes Mrs. Wallace a letter that says much about the dangers of sickness and disease prevalent in Belém at this time.[97]

Meanwhile Bates's father put him under immense emotional pressure to return even telling him how dearly his mother, who was not well at that time, longed for this.

One letter included this heartrending coercion:

> "I am happy to say your mother is greatly and I almost might say miraculously recovered her health and she now while I am writing sits in front of me as smiling and blooming as she was ten years ago and besides the medicals tell us with proper care she may live to be an old woman. [That is providing you return home.]"

---

[97] It is to be found in the Natural History Museum's Wallace Collection, reference WP 1/3/22: Pará 13th June 1851.
Dear Madam

I am very sorry to be the bearer of very bad news to yourself & family but believe it to be my duty to communicate what has happened as being the only person here nearly connected with your sons. The event we deplore is the death of your son Edward (also known as Herbert) who breathed his last here on Sunday morning last at 2 O'clock, a victim of the fatal black vomit the worst form of yellow fever.

My poor young friend had arrived from the interior about three weeks & had engaged a passage immediately in a vessel to leave for Liverpool on Friday the 6th Inst. To amuse the time until the ship sailed he had taken the same lodgings he had had with his brother in the suburbs very pleasantly situated near the forest & was very frequently at my house which was in the neighbourhood.

On the day he was taken ill we were in the city together took a cup of tea at Mr. Millers & went round to make a few small purchases. This was Monday night the 2nd Inst. On this night he was taken with a shivering & immediately fever & vomit so as to be unable to reach home, I therefore took him into a house on the road where I knew he would be as well or better attended to than at his lodgings. It happened well that he remained here as we should not have been able to have induced a medical man to go out so far to attend a patient - illness being more very prevalent in the city.

We got immediately the hotel medical advice, thinking his disease was merely constipation as it is called here but the Doctor treated him for the yellow fever & he was progressing very well on Tuesday when he committed the great improvidence of getting up and walking barefoot about a cold brick floor after mustard plasters had just been taken from his feet. The fever immediately struck inwards & black vomit declared itself early on Wednesday morning resisting all the skill of Dr. Camillio, until he died as I have already stated after suffering fearfully.

It will be more consolation to you to know that he met with the kindest attention from the English residents here especially from the Vice consul Mr. Miller who frequently visited him. I myself slept by his side four nights when I was rather alarmed by being suddenly seized with similar symptoms myself, shivering fever & vomit in rapid succession but being of lighter constitution I suppose it did not lay as firm a hold of me, I got better in four days though even now am a little weak from its effects. Poor Edward was much regretted here as being of a genial temper & a good heart, he was in a very robust state of health: he did not converse freely after being first taken but felt upset at being taken thus when on the eve of departure for England. The little property he left is in the Vice Consul's hands who will I suppose arrange accounts with Alfred - Pará is still very sickly another death from yellow fever today.

Dear Madam Yours very respectfully (signed) Henry W. Bates

There was a letter from his mother too, full of concern, 'We all wish for your society and also require your assistance in business as it is increased so very much since you left us.'[98]

From this correspondence Bates also learns that, on their return from the Isle of Wight, his family had visited Samuel Stevens in Covent Garden.

> "We had some talk with him, [his father says,] and I particularly named the desire we had that you should return home as I then stated to him my opinion that your time would be much safer and more profitably employed as a manufacturer than in your present occupation. Of course, a scientific man like Mr. Stevens would not altogether agree with me but still as I stated to him the occupation might do very well as a dernier resort to one who had tried other things and they had proved failures and instanced Mr. Wallace." [The reference to Wallace was a little unkind and could not have helped persuade Bates to return home to England. Nor could the pressure put on him have helped.]

In October he wrote once more to Edwin Brown, 'I have made up my mind to return to the interior, intending to make a short stay at Santarém, get a small canoe, and deliberately explore the River Tapajós as far as I can'.

This newfound resolution withstood the yellow fever, the death of Herbert, and all the other uncertainties. He had resolved the agony and would stay in Brazil. This was undeniably a time of crisis for we must remember that Bates was utterly alone in the wilderness. Fortunately, for Bates and for science, he resolved his crisis and set out for Santarém into the heart of the Amazons once more, returning five years later, sickly but triumphant.

Santarém 1851-1855

Today, as in Bates's time, Santarém differs from the other river settlements. The town's appearance and setting are unique on the lower Amazons. The forests surrounding the town are quite beautiful but not as dense as those found further up the river and the spacious woods around Santarém are more reminiscent of savannah. For me, it was reminiscent of the forests of

---

[98]It is to be found in the Natural History Museum's Wallace Collection, reference WP 1/3/22: Pará 13th June 1851.

West Africa in the 1950s. The greenish black water river Tapajós, a mighty stream in its own right, meanders through these hills to its confluence with the Amazon.

In Bates's time, Santarém had an air of European culture. Strict divisions had been maintained between the races, the mainly Portuguese whites formed a patrician class owning most of the trade and property. Captain Hislop, a well-known and popular resident of Santarém, helped Bates to find a suitable house, a substantial three-roomed bungalow with brick walls and stone floors, located close to the waterfront with a pretty flower garden sloping to the beach. Bates furnished it simply and soon took on a servant named José. He had previously been apprenticed to a goldsmith and turned out to be a gem in his own right remaining with Bates until he left Brazil.

Bates commenced the essential social round introducing himself to the town's notaries, judge, military commander, police commander, public prosecutor and so on down a long list thus fulfilling all expected courtesies. He obtained permission to stay in the town from the police chief, and then remained at home for the return visits by local residents confirming that he was welcome in their community.

He was astonished to find that, here in the tropics, the patricians of Santarém were so formal that they called on each other in a manner that would not have been out of place in Lisbon. Dressed in their black dress coats, regardless of the heat and humidity, they sat delicately on lacquered and gilded furniture as they exchanged courtesies. They would then withdraw in the most formal manner, retreating with frequent bows towards the exit. Bates could not match them in style but was readily accepted because the locals envied and were in awe of his scientific knowledge. It appeared that both science and geography were completely 'closed books' to most of them. He thought the people of the town felt it necessary to maintain formal standards, as they knew they were on the edge of the wilderness. All could see it in close proximity, surrounding their outpost of civilisation. Thus, they reasoned, their society would crumble if standards were lowered. Bates was astonished to find that few of the inhabitants had been more than a mile from the town by land, a reluctance motivated more by fear of the interior than anything else.

A curious little incident occurred when Bates was asked by one of his hosts 'on which side of the river was Paris to be found' meaning of course the Amazon, rather than the Seine. Bates fitted in to this community without much trouble and found the company pleasant. The climate was reasonable, the sky was usually cloudless and there was always at some time each day a breeze from the east. He particularly liked José who rapidly became more of a friend than a servant.

There were no paths except for the occasional Indian trail which few would brave, making it difficult to explore the bordering forest. There were believed to be dangerous runaway Negroes in the vicinity and it was considered unsafe to travel too close to their supposed locality. For much of his stay, Bates used the town as his base and the river as his route for exploration. Slowly he went further into the interior along a tributary of the Tapajós called the Uruará where he found more and more interesting specimens to send to Stevens. He was even critical about Wallace, saying to Stevens in January 1852 that 'Wallace cannot have worked the area so well'. Bates would set off into the interior each day accompanied by José and two boys, one Negro and one Indian, who were the bearers. They met few people in the forest but came across a curious old woman called Cecília who dwelt in the forest near to the town all year round. She fascinated Bates and, though harmless, had the reputation of being a witch. He discovered she knew a great deal about the medicinal properties of the forest plants. Bates found few animals but butterflies and insects were abundant and him occupied most of his time.

Between January and April 1852 Bates was contemplating making a trip up the Tapajós and in the middle of May hired a two masted coberta 'of six tons burden' at the cost of one shilling and two pence a day which he set about converting into a houseboat with cabin and workroom. His collecting equipment was stowed in the workroom and his important reference books arranged in the cabin. There were also his guns, ammunition, carpenter's tools, food for six months and trade goods such as looking glasses, beads, fishhooks, arrowheads, printed cottons, and locally distilled rum, all to be used for paying his way through the territory of the Tapajós. Finding a crew for the venture proved typically difficult, with little help from the local hierarchy who remained indifferent to his needs. Eventually he found two crew; one of them a young mulatto; the other, named Pinto, was familiar with the Tapajós. Recruiting them delayed the departure even more as the police would not let him travel until Pinto had paid his debts, which of course had to be settled by Bates.

At the start of the trip there was severe weather, the sails were damaged and the towed montaria had to be cast adrift and was lost. They put in to Alter do Chão, (named after the homonymous Portuguese city in the Alentejo; "Alter" being derived from the Roman name "Abelterium" given to that city), for repairs, and although Bates thought the atmosphere uninviting, he found the location - at the junction of the Tapajós and the Amazon - unrivalled in its beauty. Collecting was good there and the species different from those he had so far collected in the region of Santarém. The governor of the village was a slovenly old man by the name of Captain Thomas, a half-breed, who nevertheless helped by providing a crewmember named Manoel. Despite searching exhaustively they could not find the montaria, nor could they buy a replacement.

They left Alter do Chão and sailed along the usual impenetrable wall of unending forest but now with fine white sand beaches along the edges. These beaches were in places covered with thousands of the most dreaded of ants, the fire ant. Although they appeared to be dead, these creatures infested the region to the extent that in places it was impossible to walk on the sands.

Bates carried a letter of introduction with him to the Inspector of Indians. Called Cipriano, he lived at a place on the Paquiatuba creek, some two days up river, just beyond Pedreira. Bates searched for the governor in the flooded creeks that now stood 18 feet above the normal water level. He sighted an Indian boy who fled when he saw Bates, but was eventually coaxed out of his hiding place by Manoel. On disembarking, they eventually located the Inspector's house where they met his wife. She was a good-looking Mameluco woman according to Bates, but the Inspector was away with most of the men and Bates was unable to gain the help he desired. They returned to the boat surprised at the reserved style of reception they had received here, since he was more used to the cordiality of the other river people he had encountered.

A little further on they met a white Brazilian planter living in what appeared to be an idyllic setting. He was extremely gracious and sold them some chickens and other goods needed as provisions for the journey. His hospitality extended to inviting all the travellers to stay in his house, but Bates declined, assuming the mosquitoes would be unbearable and so remained on his boat. Pinto, who did stay ashore, returned to the boat the next morning much the worse for wear and Bates was concerned about how little he knew of the background of the crewmembers he had recruited. He sailed on to Aveiros where he established himself in a room provided by Captain Antonio, the local governor who had been forewarned of his arrival. Pinto was soon drunk once more. Bates sacked him and Manoel and the boy decided to leave at the same time, leaving Bates crewless again. However, little intervened with the routine of collecting in the woods surrounding the small settlement. Bates collected in the morning, sorted, classified, and preserved his specimens during the afternoon, studying their natural history in the evenings. It was not long before he had an enormous haul of butterflies, more than 300 different species found within a short walk of the house.

The plague of the place was the fire ant, a formidable attacker with little discrimination, a creature that had previously driven the population away, though they were just returning thinking that they would be relatively free of the pest. Biting and secreting formic acid into any wound, they are the scourge of the area and man seems to have no defence against them other than avoiding the area the ants occupy. The people of Aveiros believed they were a reincarnation of the dead *cabanos* from the 1835 rebellion, attacking humans out of sheer malice.

Captain Antonio was courteous and helpful, eventually finding two new canoe men for Bates. He took him on what turned out to be an abortive excursion to look for a particular monkey called the white cebus, but Bates acquired a parrot, which had suddenly fallen out of the sky into the river and was to remain with him for two more years. It would accompany him on all his walks perched on the head of one of the party, noisily serenading them as they pursued birds and butterflies.

On leaving, Bates sailed on until 7th August 1852 when he reached the house of the last civilised settler on the Cupari river, a hunter called João Aracu. Bates wanted to visit the Mundurucu people who were to be found nearby and Aracu was the man to take him to them. However it was to be in his time rather than Bates's so he spent the next twelve days collecting while the two crew members, Ricardo and Alberto, carved a new montaria from a tree they felled in the vicinity, which they hollowed out with chisels around a slit they had made over its entire length.

Food was scarce and every few days the men would have to stop whatever they were doing to forage for their next supplies. They fed on such things as anteaters, iguanas' eggs and turtles, but even these were in short supply. It was not that the animals were scarce, but they were well dispersed in the forest and therefore difficult to track and capture. There were, however, lots of small fish, many too small for the pot, but excellent for Bates to preserve as he did with more than forty species new to science. However, he realised that he would have difficulty preserving anything that might have gone into the cooking pot for fear this would produce a mutiny.

He shot some fine white whiskered spider monkeys and after saving their skins for collectors, even these carcasses went into the pot. According to Bates, this was the best meat he had enjoyed for a long time. It was now

19th August and Bates hoped that João Aracu would take him to an area in the interior where he would not only see Mundurucu Indians, but would also be able to capture the Hyacinth macaw. However, the location they arrived at after two days travel was in turmoil as the Mundurucu settlement had recently been attacked by roving members of a primitive food gathering tribe called the Pararuátes. The settlement consisted of scattered huts along the riverbank with walls of framework filled with mud, and thatched roofs of palm leaves, the broad eaves reaching halfway to the ground. These huts were scattered for six or seven miles along the river, invariably on tracts of level ground at the foot of wooded heights.

The Pararuátes had ransacked the plantations of the Mundurucu who had pursued them into the forests seeking revenge. The Tushaua, or headman, of the Mundurucu had led the pursuit with thirty of his warriors, armed with guns, bows-and-arrows and javelins. For two days, without success, they hunted them through the forest, and now were reclining in hammocks outside the house of the chief, discussing the chase and their failure. The first strange figure to catch Bates's eye was a tall man, whose face, shoulders and breast were tattooed all over in a crossbar pattern. The next had an even more extraordinary appearance, with a semicircular patch in the middle of his face, covering the bottom of the nose and mouth; crossed lines on his back and breast, and stripes up his arms and legs.

To complete the picture of a savage community, completely naked women were making farinha. As Bates approached them the women fled into the shadows at the sight of this ghostly apparition, a whiskered bespectacled white man, with nets, guns and other paraphernalia of the collector around his person.

The Tushaua fascinated Bates. In the first place, he spoke Portuguese as he had visited Belém. He had no tattoos and was dressed in a shirt and trousers with nothing about him of the savage. Bates was intrigued and they quickly got into conversation. Sometime later, he was presented with the only trophy the Mundurucu had collected in the pursuit, a small necklace of scarlet beans, which had fallen from the fleeing Pararuátes. Bates was flattered by this gesture and went back to his canoe to collect his *Knight's Pictorial Museum of Animated Nature*, a two-volume tome that formed part of his travelling reference library. This caused the Tushaua much amusement and interest, enough for him to gather his many wives around to share the excitement engendered by the illustrations in the book. Bates noted that one of the wives was a fine-looking girl dressed only with strings of blue beads. Excitement continued to grow and soon many of the Munduruců were gathered round and at least sixty bare breasted women and naked children, relentless in their curiosity, surrounded Bates while they studied and jabbered about every illustration the books contained.

These people were noted for the fine feather work of their headdress which was used at the time of celebrations and supposed to have supernatural properties. Enchanted by this extraordinary traveller in their midst with his book of exquisite drawings they allowed him to purchase two items of feather work, a most unusual occurrence in those days.

Bates also witnessed the tribe's medicine man in action treating a child with severe headaches by blowing leaf smoke on the child's head, then sucking the flesh, afterwards pulling from his mouth a worm he claimed to have been the cause of the child's discomfort. João Aracu retrieved the worm for closer inspection and discovered that it was in fact a root of some description that the medicine man had slipped into his mouth beforehand. Bates watched this whole affair with scepticism commenting that he thought the practice was bogus and the practitioner a quack.

This part of the journey did much to restore Bates's feeling for the Indians, whom he now regarded as simple, gentle, well-mannered people with an abundance of cheerfulness and frankness, rather different from the images he had previously held. He found their beliefs disarmingly simple and naïve and according to Woodcock, found it difficult to get at the Munduruců philosophy on subjects that required a little abstract thought. He dismissed their religious beliefs as a faith in hobgoblins, and decided that their life was monotonous and dull.

Early the next morning they left to continue their journey to the Cupari falls where Bates hoped to find the hyacinth macaw. The air was dank, the forest tumbled down to the waterfront and there was a gloom caused not only by the early morning mist but also by the slow rising of the sun above the canopy far above them. The falls were reached later in the day and marked the end of navigation for a craft of the size they were using. There were two cataracts so Bates divided the party into two, José, João, and one of the Indians searching the upper cataract 15 miles upstream whilst he and the remaining Indian stayed close to the lower falls. Within four days he had his prize, a haul of four hyacinth macaws and a new species of howler monkey, *Mycetes seniclus*. Bates, now serenaded by a chorus of night forest noises, had to prepare his specimens, each macaw taking as much as three hours to preserve.

> "Then, almost imperceptibly, the frogs begin their nightlong racket. The crickets and cicadas join in and there are other nocturnes in monotones from creatures unknown. There are notes so profound, but constant, that they startle you...and there are stridulations so attenuated that they shrill beyond reach...Then the howler monkeys begin, and as they do they silence the rest of the cacophony but make a din that out decibels all the other noises of the forest put together."

Bates was now using the boat as a base and making regular forays ashore into the woods; however the loneliness and isolation began to tell not only on him but also on his men. To an extent they were afraid of the Pararuátes who were still thought to be somewhere in the area and would definitely have been dangerous to encounter after their mauling by the Mundurucú. This feeling applied particularly to the party further upstream. Therefore, pleased with the hyacinth macaws, Bates turned back on 26th August to return to João Aracu's sítio for a two-week stay, during which time he collected many beautiful butterflies and other insects. The heat, always having to push his crew to achieve anything, the bad food and such matters were beginning to tell on Bates, so he was delighted to reach the open waters of the Tapajós River once more. They sailed on to Averyos only to find the place fever-ridden, so he settled his accounts with Captain Antonio and hurried on.

The weather now took its turn at making life uncomfortable and dangerous as a squall sprang up and tossed the boat around in shallow waters. Bates became concerned that, as he tried to reach the shore and the residence of Inspector Cypriano, he could lose his valuable collections, either by their getting thoroughly wet or being washed overboard in the storm. After this alarming experience he decided to go no further on the river until he could find an experienced pilot to deal with the unusually shallow water. Inspector Cypriano obliged by lending him one of his men. Named Angelo Custodio (which was rather apt for these circumstances as it translates as *guardian angel*), he safely guided them back into Santarém on 7th October 1852. To his surprise and disappointment, Bates discovered that Wallace had passed through Santarém on his way back to Belém (and eventually to England) and had wanted to see Bates if only to find out the circumstances of his brother Herbert's death. One can only speculate on Bates's mood as he struggled with illness and fatigue. Here was his one-time travelling companion returning to England at a time when he, Bates, was sick and tired of being a travelling naturalist in the remote forests of Amazonia. He must now have felt he was alone in the wilderness even though he had not recently been travelling with Wallace. When a boat left Santarém a little while later, on the 18th October, he still had not gathered the strength to pack up the Tapajós specimens and send them on to Belém and ultimately to Stevens. Instead he wrote to Stevens on 22nd November, saying:

> "I wrote to you soon after my return, saying how unwell I was at the time; I am sorry to say I have been still worse, and have suffered severely from the fatigue of the voyage and the unhealthiness of the Tapajós [nearly all the branch rivers here are pestilential]; but still I have had no acute diseases, merely bilious attacks, head-aches, prostration of strength...medicine only made me worse; so I put myself on a rigid diet, took gentle exercise, avoided close application, and exposure to the sun."[99]

Bates had become aware of the unpredictability of hiring boats for journeys of complete speculation. During the last one on the Capari, he had spent so much on provisions that his butterfly collecting on the journey would have had to amount to at least 4,000 saleable specimens at 3 pence per specimen, i.e. (£50), for him to break even. He decided that this was the last journey he would make of this kind. The idea of returning to Tefé attracted him but there were no boats scheduled from Santarém to Tefé in the near future. However, there was a rumour that a steamer was about to make the journey from Belém to Peru. If this turned out to be true, he decided he would make the journey. In fact it was to be two more years before he was able to leave Santarém for his second visit to Tefé. In between, his health had improved, so by March 1853 he was telling Stevens he was not prepared to leave Brazil without first exploring Tefé and beyond.

In June 1853, he was asking Stevens to send him periodicals to read so that he would 'feel less an exile from intellectual society'. Anything would do, not necessarily new, second hand would be as welcome and the subject was less important than readability, although he expressed a desire to have the *Athenaeum* and *The Zoologist* if possible. Bates was later to comment that he read the periodicals he received which included the *Atheneaeum* many

---

[99] Letter to Stevens, 22nd November 1849, *Zoologist* 8 (1850) 2665.

times over during his loneliness and intellectual starvation. First, he would read the advertisements, then put the magazine away to take it out on other occasions to read an article, then put it away again so that he could ration access to what was eventually to be the only contact he had with anything remotely stimulating.

In the meantime Stevens was encouraging him to travel once more on the Tapajós to collect more specimens and, as his health improved, his plans clarified. He determined he would travel to Tefé as soon as the opportunity presented itself, but in the intervening time he would again go to Alter do Chăo. In the end, he secured a passage for the short distance arriving there on 27th August 1853. He found the place idyllic, despite the limitations of the people he knew and met. He would set out in the evening to lie on the beach and in the cool moonlight dream about English weather. By mid December he was once more in Santarém where he intended to remain for a few weeks in order to enjoy the town's civilised comforts. After a while he moved on to Villa Nova. He found the collecting to be sparse, but there he met the French naturalist, Emile Carrey who was travelling back through Vila Nova from Peru to Belém with a collection of magnificent butterfly specimens which stirred envy in Bates. He now wanted nothing more than to reach Tefé to search out treasures like those Carrey had shown him. Villa Nova was not a success as Bates was once more overcome by fatigue which seemed to be brought on by bad food, hard work and loneliness. At about this time he heard from Wallace who was embarking for the Malay Archipelago in order to continue his collecting and of the disastrous fire that had destroyed all of Wallace's Amazonian collections.

On 5th June 1855, more than two years later than he first anticipated, he set out for Manaus in order to connect with a passage to Tefé. José accompanied him on this journey to the upper Amazons and it would be almost four more years before they would return.

The Second Journey to Tefé and Beyond 1855-1859

The introduction of the government steamer service on the Amazons in 1853 reduced travelling times considerably. When Bates first journeyed to Tefé in 1849-1850, his starting point had been Óbidos and the journey had taken more than three months. Bates now realised that beginning his journey at Santarém further to the east, the journey would take only a few days including a six-day stop in Manaus. The steamer was captained by Major Estalano of Tefé who, in 1874, met with Charles Barrington Brown and

William Lidstone on their journey up the Amazon on behalf of the Amazon Steam Navigation Company of London. Estalano related the story of Bates's journey to these two travellers. Not only was the journey faster but the quality of travel was better. There were cabins and a deck where passengers could sit and enjoy the scenery and professional cooks on board preparing reasonable meals. Sailing 'in the lap of luxury,' Bates reached Tefé on 19th June 1855.

He stayed there until 17th March 1859 when he commenced his descent to Belém on the journey home to Leicester and England. In 1855, the floodwaters were just beginning to ebb, the ground starting to dry out and pathways and forests were becoming accessible once more. This time Bates made a colossal haul of specimens, no less than 550 different species of butterfly, and 7,000 insects in all, of which 3,000 were found to be new to science. These are astonishing numbers. Today, most entomologists would not experience anything on this scale during an entire lifetime of collecting. Bates noted in September 1855 that he found four or five new species each day:

> "...those who know a little of Entomology will be able to form some idea of the riches of the place in this department, when I mention that eighteen species of true Papilio [the swallow-tail genus] were found within ten minutes' walk of my house. No fact could speak more plainly for the surpassing exuberance of the vegetation, the varied nature of the land, the perennial warmth and humidity of the climate. But no description can convey an adequate notion of the beauty and diversity in form and colour of this class of insects in the neighbourhood of Ega. I paid special attention to them, having found that this tribe was better adapted than almost any other group of animals or plants to furnish facts in illustration of the modifications which all species undergo in nature, under changed local conditions. This accidental superiority owes partly to the simplicity and distinctness of the specific character of the insects, and partly to the facility with which very copious series of specimens can be collected and placed side by side for comparison. The distinctness of the specific characters is due probably to the fact that all the superficial signs of change in the organisation are exaggerated, and made unusually plain by affecting the framework, shape, and colour of the wings, which, as many anatomists believe, are magnified extensions of the skin around the breathing orifices of the thorax of

> the insects. These expansions are clothed with minute feathers or scales, coloured in regular patterns, which vary in accordance with the slightest change in the conditions to which the species are exposed. It may be said, therefore, that on these expanded membranes Nature writes, as on a tablet, the story of the modifications of species, so truly do all changes of the organisation register themselves thereon. Moreover, the same colour-patterns of the wings generally show, with great regularity, the degrees of blood relationship of the species. As the laws of Nature must be the same for all beings, the conclusions furnished by this group of insects must be applicable to the whole organic world; therefore, the study of butterflies-- creatures selected as the types of airiness and frivolity-- instead of being despised, will some day be valued as one of the most important branches of Biological science."[100]

In November 1856, he travelled by steamer a further 240 miles upriver to Tonantins and Fonte Boa, where he remained for about three months. In September 1857, he visited São Paulo de Olivença, another 150 miles beyond Tonantins and close to the Peruvian border where he stayed for about five months. The remainder of his time was spent near Tefé, collecting specimens on the numerous walks that were possible near the town. Bates found that this was indeed the richest collecting ground he had discovered anywhere on the Amazons.[101]

Bates soon settled in Tefé, meeting old friends from his previous stay, and relaxing into the local dress style in clogs and check shirt. He took part in village life even acting as godfather at the ceremony of baptism for some Indian children. He says nothing of the view the white population of the town took of these associations but once again, he was probably seen as *going native*. For a long time during his time in South America, Tefé was the focus of his life and there is no doubt Bates felt romantic about the place. Friends he had made on his earlier visit were to die during this second visit and Bates witnessed their burials, so there was no way he could have avoided attachment to both the people and the place even had he wished to do so. He had known the place for long enough to see young people grow up, attend their weddings and see their offspring.

---

100 One should remember that Bates was well aware that even Darwin considered collecting butterflies a frivolous activity.

101 The Proceedings of Natural History Collectors in Foreign Countries. *Zoologist* 15: 5557-5559.

The place had changed. On his first visit it was little more than a village, but by 1855 it had grown into a town. Admittedly it had not altered in size that much, but now there were resident judges and bureaucrats of all types. Every two months a boat would arrive from the coast bringing with it all the appurtenances of style previously missing in such a remote location. The loss of its earlier isolation had penalties as Bates discovered. For instance, the price of turtles had risen ten fold over the period of his absence. The staple food in Tefé at this time was turtle but Bates grew so tired of the unchanging diet he eventually said he would rather go hungry than eat another turtle. For a full two years he did not eat wheaten bread, using tapioca soaked in coffee as a substitute. This poor diet was undoubtedly one of the causes of his physical deterioration whilst there.

The English Monkey

The white uakari, sometimes called the English monkey after its peculiar complexion, is found only in the várzea-flooded forests enclosed by the Solimões and Japura rivers and the Auati-Parana canal near Tefé. Bates first described it and, as recently as the 1980s this was the only description of the uakari, *Cacajao calvus ulrus*, in its natural habitat. Bates made extensive studies of the flora and fauna of the seasonally flooded forests that extend along the Solimões and its tributaries. With his departure, the white uakari went unstudied for a hundred and fifty years, when, in March 1983, the late Jose Márcio Ayres, a young biologist from the state of Belém, set out from Tefé in an old boat to look for the mysterious scarlet faced monkey. Abandoning several field sites as too remote, others because of the ferocity of biting insects, he finally located the Mamiraua lake system where the Mamiraua Sustainable Development Reserve now protects this strange monkey. I was thrilled in my turn to see this extraordinary monkey in these same forests.[102]

Becoming Accustomed to the Place

The loss of isolation changed the very nature of the place. Travellers took advantage of better communications and passed through the town on their way from Peru to the coast and vice versa. Some of European origin found

[102] Ayres, José Márcio, *As Várzeas do Mamirauá.*, Sociedade Civil Mamirauá, 1995. Padoch, Christine; Ayres, J Márcio, Pinedo-Vasquez, Miguel, Henderson, Andrew (Ed), *Várzea. Diversity, Development, and Conservation of Amazonia's Whitewater Floodplains. Advances in Economic Botany.* Volume 13, (New York: The New York Botanical Garden Press).

it an attractive place and settled there marrying local Indian or mameluco girls.

In 1827, Edwards, still in his early twenties commented on these girls:

> "Some of our party went on shore to look up old acquaintances. I remained on board, preferring to make observations by daylight. It was late before the noise in the town subsided, what with muskets and rockets, singing and fiddling so late, that I must have been dreaming hours before; but the first thing that awoke me in the morning was a splashing, and laughing, and screaming all around the galliota, where the sex par excellence, was washing away the fatigues of the dance in a manner: to rival a school of mermaids. And these Indian girls, with their long floating hair and merry laugh... darting through the surf like ...sea-nymphs."[103]

One of these travellers, a Frenchman, started a tailors shop in the town and visited Bates on a regular basis, usually arriving with a monkey on his shoulder. Another friend he found settled in Tefé was his old acquaintance João de Cunha Correia, with whom he had voyaged from Belém to Óbidos in 1849. Hauxwell, the ornithologist, also appeared in Tefé from time to time and would always stay with Bates as he prepared for another mission into the interior with his faithful Indian hunters in search of new bird specimens.

Bates revelled in the company of these intelligent men and no longer felt so isolated. However he craved their company when they were not there. He became much more domesticated, gathering a household around himself led by José, consisting of both Indians and animals, together with a pet toucan.

> "One day...while walking along the principal pathway in the woods near Tefé, I saw one of these Toucans seated gravely on a low branch close to the road, and had no difficulty in seizing it with my hand. It turned out to be a runaway pet bird; no one, however, came to own it, although I kept it in my house for several months. The bird was in a half-starved and sickly condition, but after a few days of good living it recovered health and spirits, and became one of the most amusing pets imaginable. Many excellent accounts of the habits of tame

[103] See Edwards, William H, *A Voyage up the River Amazon, Including a Residence at Pará*, (London: John Murray, 1847).

Toucans have been published, and therefore, I need not describe them in detail, but I do not recollect to have seen any notice of their intelligence and confiding disposition under domestication, in which qualities my pet seemed to be almost equal to parrots. I allowed Tocano to go free about the house, contrary to my usual practice with pet animals, he never, however, mounted my working- table after a smart correction which he received the first time he did it. He used to sleep on the top of a box in a corner of the room, in the usual position of these birds, namely, with the long tail laid right over on the back, and the beak thrust underneath the wing. He ate of everything that we eat; beef, turtle, fish, farinha, fruit, and was a constant attendant at our table--a cloth spread on a mat. His appetite was most ravenous, and his powers of digestion quite wonderful. He got to know the meal hours to a nicety, and we found it very difficult, after the first week or two, to keep him away from the dining room, where he had become very impudent and troublesome. We tried to shut him out by enclosing him in the backyard, which was separated by a high fence from the street on which our front door opened, but he used to climb the fence and hop round by a long circuit to the dining-room, making his appearance with the greatest punctuality as the meal was placed on the table. He acquired the habit, afterwards, of rambling about the street near our house, and one day he was stolen, so we gave him up for lost. But two days afterwards he stepped through the open doorway at dinner hour, with his old gait, and sly magpie-like expression, having escaped from the house where he had been guarded by the person who had stolen him, and which was situated at the further end of the village."

During this time Bates also acquired two Indian children. The circumstances surrounding their arrival in his household are both unusual and interesting. He says that in a compassionate moment he purchased the children through a trader who operated among tribes practising slavery in the more remote areas but who were always willing to sell on children should an opportunity to do so arise. There were two of them, a boy of about 12 or 13 and a younger girl. Bates called the boy Sebastian and it is possible he came to Bates in this way. The little girl, called Oria, may be a different story, for it seems that she was delivered to his door during one night of inhospitable weather.

"My assistant Jose, in the last year of my residence at Ega, [1859] 'resgatou' [ransomed, the euphemism in use for purchased] two Indian children, a boy and a girl, through a Japura trader. The boy was about twelve years of age, and of an unusually dark colour of skin; he had, in fact, the tint of a Cafuzo, the offspring of Indian and Negro. It was thought he had belonged to some perfectly wild and homeless tribe, similar to the Pardrauates of the Tapajós, of which there are several in different parts of the interior of South America. His face was of regular, oval shape, but his glistening black eyes had a wary, distrustful expression, like that of a wild animal; and his hands and feet were small and delicately formed. Soon after his arrival, finding that none of the Indian boys and girls in the houses of our neighbours understood his language, he became sulky and reserved; not a word could be got from him until many weeks afterwards, when he suddenly broke out with complete phrases of Portuguese. He was ill of swollen liver and spleen, the result of intermittent fever, for a long time after coming into our hands. We found it difficult to cure him, owing to his almost invincible habit of eating earth, baked clay, pitch, wax, and other similar substances. Very many children on the upper parts of the Amazons have this strange habit; not only Indians, but Negroes and Whites. It is not, therefore, peculiar to the famous Otomacs of the Orinoco, described by Humboldt, or to Indians at all, and seems to originate in a morbid craving, the result of a meagre diet of fish, wild fruits, and mandioca meal. We gave our little savage the name of Sebastian. The use of these Indian children is to fill water-jars from the river, gather firewood in the forest, cook, assist in paddling the montaria in excursions, and so forth. Sebastian was often my companion in the woods, where he was very useful in finding the small birds I shot, which sometimes fell in the thickets amongst confused masses of fallen branches and dead leaves."

According to Bates, Sebastian:

"...was wonderfully expert at catching lizards with his hands, and at climbing. The smoothest stems of palm-trees offered little difficulty to him: he would gather a few lengths of tough flexible lianas; tie them in a short endless band to support his feet with, in embracing the slippery shaft, and then mount upwards by a succession of slight jerks. It was very amusing, during the first few weeks, to witness the

> glee and pride with which he would bring to me the bunches of fruit he had gathered from almost in accessible trees. He avoided the company of boys of his own race, and was evidently proud of being the servant of a real white man. We brought him down with us to Belém; but he showed no emotion at any of the strange sights of the capital: the steam-vessels, large ships and houses, horses and carriages, the pomp of church ceremonies, and so forth. In this he exhibited the usual dullness of feeling and poverty of thought of the Indian; he had, nevertheless, very keen perceptions, and was quick at learning any mechanical art. Jose, who had resumed, some time before I left the country, his old trade of goldsmith, made him his apprentice, and he made very rapid progress; for after about three months' teaching he came to me one day with radiant countenance, and showed me a gold ring of his own making."

"Ecui encommenda"[104]

The story of the children becomes confusing as Bates writes about the little girl, 'who came with a second batch of children a month or two after Sebastian'. He is not consistent on this subject, as earlier he says the children were ransomed together. He later continues:

> "She was brought to our house one night in the wet season, when the rain was pouring in torrents, thin and haggard, drenched, and shivering with ague. An old Indian, who brought her to the door, said briefly, ecui encommenda and went away."

Bates admits:

> "There was very little of the savage in her appearance, and she was of a much lighter colour than the boy. We found she was of the Mirahné tribe, all of whom are distinguished by a slit, cut in the middle of each wing of the nose, in which they wear on their holiday occasions a large button made of pearly river-shell. We took the greatest care of our little patient; had the best nurses in the town, fomented her daily, gave her quinine and the most nourishing food; but it was all of no avail: she sank rapidly; her liver was enormously

[104] *Here is your little parcel.*

swollen, and almost as hard to the touch as stone. We had an old woman of the same tribe to attend her, who explained what she said to us. She often begged to be taken to the river to bathe; asked for fruit, or coveted articles she saw in the room for playthings."

The Mirahné tribe, thought to be cannibals, is from an area to the north west of Tefé. Bates, finding the child sick, immediately found nurses to look after her and put all his amateur knowledge of medicine into her care, skills he had acquired on his long journey. He seemed moved by a compassion that verged on agonised love for this little girl. However, despite all his efforts, this dying child from another world was beyond his saving.

Bates was in Tefé in 1850 and now again eight years later in 1858 so the age of the child corresponded with his previous connection with the area. Could she have been his child, the result of a liaison in 1850 with a local girl? If this was so maybe the old woman who brought her to Bates knew this and delivered the child to him in her hour of need? If anyone could do something it might be the white traveller and here he was once more in Tefé. Bates comments in his book are worth noting in full:

"There was something uncommonly pleasing in her ways, and quite unlike anything I had yet seen in Indians. Instead of being dull and taciturn, she was always smiling and full of talk...The last week or two she could not rise from the bed we had made for her in a dry corner of the room; when she wanted lifting, which was very often, she would allow no one to help her but me, calling me by the name of 'Cariwa' [white man], the only word of Tupi she seemed to know. It was inexpressibly touching to hear her, as she lay, repeating by the hour the verses which she had been taught to recite with her companions in her native village; a few sentences repeated over and over again with rhythmic accent, and relating to objects and incidents connected with the wild life of her tribe."

Whatever the truth Bates appeared to be emotionally involved with this child even though her stay with him was brief. His actions seem to have been totally out of character, filled with sentiment and oblivious to the reaction it caused among those around him.

Bates was so concerned by Oria's suffering that he fetched his friend the Padre to baptize her, and when she died he irritated 'the big people of Tefé' by insisting that she was buried with all the ritual usually only accorded to a white Christian child:

> "We had the corpse clothed in a robe of fine calico, crossed her hands on her breast over a 'palma' of flowers, and made a crown of flowers for her hair."[105]

Plate 9.
An Old Grave similar to Oria's in Tefé

Life had to go on

There was an interesting altercation with a toucan at Tefé:

> "I had shot one from a rather high tree in a dark glen in the forest, and leaving my gun leaning against a tree-trunk in the pathway, went into the thicket where the bird had fallen, to secure my booty. It was only wounded, and on my attempting to seize it, it set up a loud scream. In an instant, as if by magic, the shady nook seemed alive

[105] We will never know the truth, but there certainly was a special relationship between Bates and the child Oria. I searched for any trace of this child in Tefé during my visit in 2005 and the Bishop of Tefé confirmed there were no records of Indian baptisms before 1888 when slavery was abolished. Because of this edict, once the slaves were free, there was no record of the outcome of any form of coercion during bondage. The cemetery where she was buried was the only one in Tefé and it is still there. Called the Cemetario Catedral Da Saudade, it was remodeled in 1893, many of the poorer graves being worked over as new burial ground. In the 1950s it was remodeled once more, this time being considerably reduced in size to provide building land for the now sprawling town of Tefé. Sr. José Rocka, the head groundkeeper at the cemetery knew nothing of the whereabouts of the grave but showed me some unmarked and old ones that could have been there in Oria's time commenting that hers would have been similar to these ones.

with these birds, although there were certainly none visible when I entered the thicket. They descended towards me, hopping from bough to bough, some of them swinging on the loops and cables of woody lianas, and all croaking and fluttering their wings like so many furies. Had I had a long stick in my hand, I could have knocked several of them over. After killing the wounded one I rushed out to fetch my gun; but the screaming of their companion having ceased, they remounted the trees, and before I could reload every one of them had disappeared."[106]

Plate 10.
The incident with the toucans

It was in Tefé that Bates decided mimicry was a certainty and it was here that he reached his lowest point with loneliness and fatigue.

Woodcock says:

[106] Amabel Williams-Ellis republished this engraving taken from Bates's book (*Darwin's Moon: A Biography of Alfred Russel Wallace*. Blackie, London & Glasgow. 1966), with the protagonist's features accentuated to make him look like Wallace, but it is clear from the original text that it was Bates who had the experience. In both cases the artist has depicted a gun in Bates's hand which was not the case when the toucans were croaking at him at close range. Helena Cronin again republished Williams-Ellis's altered picture as Wallace (*The Ant and the Peacock*. Cambridge University Press, Cambridge. 1991). This is clearly a case of author's licence but it is also an instance where events are distorted sufficiently to overshadow Bates.

> "...from his scientific discoveries, he was still unable to accept ultimate absorption into a world that lacked the varied excitement of European life." [107]

Writing to brother Frederick in September 1855, telling him of the joy of finding beautiful unique specimens of new species of butterflies, Bates went on to say:

> "When you consider the great pleasure there is in this, and at the same time the liberty."

There must have been many times when Bates envied the single-minded Frenchmen and Italians who were able to settle down in Tefé without any apparent qualms about remaining there for the rest of their lives, severed from the complex civilization they had abandoned. Such a complete abandonment he could never attain. The very sharpness of observation and intelligence which made him a good naturalist and a good writer also made him incapable of accepting finally the almost wholly physical happiness of the river and its gypsy life. Even when he weighed the freedom of this life against the existence he could expect in Leicester as a hosier's clerk and threw in the satisfactions he gained independence of this kind of life,-the tolerable good living (turtle, fresh fish, game, fowls, the suavity of the climate,

> "...you will readily understand why I am disinclined to return to the slavery of English mercantile life. One great privation, however, I suffer and feel acutely the want of frequent receipt of letters, books, newspapers and magazines. Since May last I have not received one line of letter or page of European literature! The stirring news of war [Crimea] I get only through the miserable, brief notices in Brazilian newspapers." [108]

It was to be a full eight months, May 1856, before he was to receive the books for which he hungered. 'It was a happy day for me,' he told Frederick when they arrived. 'You, in England, in the midst of books and intellectual treasures, in fact, bored by their profusion cannot form any idea of the

[107] Woodcock, George, *Henry Walter Bates: Naturalist of the Amazons.* (London: Faber & Faber, 1969)
[108] Lee, Monica, *300 Year Journey: Leicester Naturalist Henry Walter Bates, F.R.S., and his family, 1665-1985.* (Leicester: Castle Printers. 1985)

luxury of receiving, in these savage solitudes, such a parcel.' Later on, books or magazines arrived more regularly, so that two or at most four months would elapse between batches of reading matter, yet he still had to be economical in his reading for fear he would finish all and be left utterly destitute before the next boat arrived. Even today it is not an easy place to reach.

Bates found that the sense of intellectual deprivation increased instead of being deadened by time, and the contemplation of Nature alone offered little compensation.

On 7th November, after Bates had written to Wallace to praise him on his paper in *The Annals and Magazine of Natural History*, he boarded the steamer *Tabatinga* for his journey to Tonantins, a small village in Indian country, about two hundred and forty miles upriver from Tefé. Learning from previous experience, he took with him all the provisions he might need for a three-month journey, anticipating the small riverside settlements he visited would not be able to provide the basics for day-to-day living. This meant he had to carry items such as cooking utensils, crockery, and hammocks. He also had an enormous mosquito tent, eight feet long and five feet broad, made of coarse calico with sleeves at each end through which to pass the cords of a hammock, as he thought the places he proposed visiting could be overrun with mosquitoes.

The journey took four days in a boat under the command of Lieutenant Nunes Mello Cardozo, an officer of the Brazilian navy, who took an interest in Bates and his activities, and became rather more than an acquaintance. Nunes was keen on discipline which was reflected in the manner he ran his boat. The passengers were a mixed lot, including four Frenchmen bound for Moyabamba on a trading mission, a German and a Moroccan Jew going to set up shop in Peru. There were also a goodly number of Peruvians dressed like Americans in denim trousers and panama hats who came from the Andean foothills, had been to Belém with their cargoes of Panama hats, and were now returning to Peru with heavy loads of English hardware and crockery.

The boat's destination, for Bates at least, was the small commune of Tonantins, a place in the wilderness, far off the beaten path, and hidden in the forests past Tefé. This village of twenty primitive mud and palm huts took its name from the slow-moving, dark, sluggish, tributary of the Amazons on which it was located. Its streets were choked with weeds and so muddy when it rained as to be almost impassable. The forest closed in on the place intimately, swamps surrounded it. The atmosphere was 'close, warm and reeking' and the humming and chirping of birds and insects never ceased. Plovers and herons haunted the scrap of scrubby open ground near the village, and opportunistic caimans floated lazily on the river in front of the houses.

Bates went ashore with José, taking all their baggage in an Indian canoe, and immediately presented himself to the only official in the town, a mameluco director of Indians named Paulo Bitancourt. Senhor Bitancourt had been warned of Bates's arrival and had generously arranged a house for him. It was a basic one room dwelling which was dirty and damp with all the tell tale marks of termite invasions. There were dirty calico blinds on the windows, the usual method used in Indian dwellings to keep out the biting pium flies, which abounded in the area, gathering in clouds as thick as smoke wherever the conditions were dank. However, it was one of the better houses in the village and Bates was grateful it had been made available to him. He stayed there for nineteen days and despite his innovative mosquito proof tent in which he could live almost entirely, he was eventually driven away not only by the persistent flies and merciless mosquitoes, but also by the complete lack of fresh food, something that did not apparently concern the native inhabitants.

He wrote to his brother Frederick:

> "I cannot stay long, for the immense number of insects pests (clouds of piums by day and mosquitoes by night), added to positive hunger (for next to nothing is to be had to eat), are beyond my powers of endurance."[109]

So awful was his intense discomfort that when a small trading schooner came into view Bates obtained passage for himself, José and all their equipment to the village of Fonte Boa - part way between Tonantins and Tefé. Francisco Raiol, a young man already known to Bates, captained the schooner which was more or less empty, having unloaded its goods from the east further up the river. He told Bates that there were plenty of mosquitoes in Fonte Boa but no piums, so half of each twenty-four hours would be tolerable as mosquitoes were generally nocturnal.

When reminiscing about his stay in Tonantins, Bates had to conclude that he had done rather well. He had made a fine collection of insects, more

[109] Lee, Monica, *300 Year Journey: Leicester Naturalist Henry Walter Bates, F.R.S., and his family, 1665-1985*. (Leicester: Castle Printers. 1985).

than three hundred species, of which about one in four was subsequently found to be new to science. Collecting that number in fourteen days was a remarkable haul. Bates was still the peripatetic collector even at this late stage in his journeys, driven by the need to catch as many fine specimens as was possible that could be converted into cash. This was something Wallace was to severely criticise in one of his letters to him. Bates found two unknown birds and a new species of monkey reconfirming his view of the remarkable local variability of the animal population on the Amazons. One of these new captures pleased him immensely for sheer glory. It was a butterfly he caught after a two-hour chase as it flitted just out of his reach from tree trunk to tree trunk in the most tantalising way. He described it with all the aesthetic enjoyment that so often enters into the behavior of the entomologist, passionately dwelling on the rich intense blue of the upper wings, varying in shade with the movement of light, and heightened with a broad curved stripe of orange on each wing, known today as *Catagramma excelsior.*

Bates was out in the forest following another overgrown track, searching for butterflies and other interesting insects when the forest began to open up and he found himself on a more clearly defined track. He suddenly stumbled across a tribe of Indians more primitive than any other he had yet encountered. They were the Caishanas, only half civilized, the adjacent forest being the territory of this secretive and peaceful clan of hunting people. They lived in huts hidden in the forest, but not in communities. It was only by chance that he discovered one of the Indians because the modest path, bordered by elegant *Lycopodiums* was blocked in many places by the trunks of felled trees, looking as if they were set up to discourage intruders. In due course Bates reached a small clearing beside a brook. On the edge of the clearing stood a conical hut with a low doorway, and beside it, an open shed built of split palm-stems.

> "Two or three dark-skinned children, with a man and woman, were in the shed; but immediately on espying me, all of them ran to the hut, bolting through the little doorway like so many wild animals scared into their burrows. A few moments after, the man put his head out with a look of great distrust; but on my making the friendliest gestures I could think of, he came forth with the children. They were all smeared with black mud and paint; the only clothing of the elders was a kind of apron made of the inner bark of the

sapucaya-tree, and the savage aspect of the man was heightened by his hair hanging over his forehead to the eyes."

Bates was now visiting an area where many small tribal groups with their own languages and customs dwelt. He approached the indigenous people cautiously and it was not long before children came out of the shadows near their hut to run around as if helping this strange and surprising intruder in the eccentric practice of capturing butterflies in his large white net.

Bates was to learn a great deal about these extraordinarily primitive people, who appeared to have lived from time immemorial at the simplest of social levels in these forests. They had neither chiefs nor medicine men. There was no obvious hierarchical system in their culture, and as far as Bates could see, they had remained consistently opposed to any attempt to convert them to Christianity by any occasional missionary who passed their way. They had blowguns with poisoned arrowheads and used them for hunting so successfully they had slaughtered all available game for many miles around their home, but the men did not wage war to secure new hunting grounds, they simply went further afield. They had no masked dances, and if they did practise ceremonies to pacify or appease the spirits, nobody had ever seen them.

From time to time, they gathered as many as they could for a kind of celebration in which the people from the surrounding scattered huts would drink weak beer made from fermented mandioca, maize or bananas. Bates discovered that these affairs never turned into the drunken orgies lasting several days and nights that he had so often seen elsewhere.

Previously, when he had been with the Juris, he watched their slow repetitive stamping dances to the accompaniment of drums made of hollow logs and beaten with the hands. The Caishanas had no such dances, and the only activity that suggested cultural display was the practice of the men who would lie in their ragged hammocks and play for hours on a kind of Pan's pipe made from the hollow stems of arrow grass. Seen in the abstract it might appear to have been an idyllic existence; in reality it was quite the opposite for the Caishanas were dirty, hungry, and forever frightened of their more belligerent neighbours.

When he left Tonantins at the end of November with Francisco Raiol, Bates encountered other remote settlements on his way to Fonte Boa. For four days they slowly made their way down river until they reached the junction with a tributary called the Jutahi, a minor river in Amazonian terms but a major waterway in European terms being half a mile wide some distance from the confluence. Here they anchored inside the mouth of the Sapo, a tributary of the Jutahi, while Senhor Raiol's men went to collect a cargo of salt fish that was awaiting them in another tributary further upstream.

Bates and José took a canoe to follow the constricted, dark stream of the Sapo. Here and there, hidden by the luxuriant forest, they saw houses of the Maraua tribe, their presence only indicated by the canoes tied up in the small coves that served as ports. Some twenty miles further up stream they reached the last Maraua houses where they landed, finding several families of Indians living in two large quadrangular buildings. Each was communal, open at the sides but in places partly enclosed by low mud walls that formed small internal chambers for individual families.

The Marauas who lived there, like the Caishanas, did not perform the tattooing usually found among the more refined Amazonian tribes, but most of the men had large holes in their ear-lobes which held round plugs of wood, and smaller holes drilled through their lips for the same purpose. Bates noticed many of the Indians had black blotches on their skin, particularly on their faces; one old man had a completely blackened face looking like lead, but though obviously an illness the cause remained a mystery to Bates.

These people had none of the shy evasiveness of the Caishanas and were engaging to the extent that one of the young men, who apart from being a good specimen of manhood and nearly six feet tall, took a particular fancy to Bates. He clowned for him, putting little white sticks into the holes bored in his lips 'and then twisting his mouth about and going through a pantomime to represent defiance in the presence of an enemy'. Possibly the sticks were mimicking Bates's beard.

Bates was amused by all this but was surprised when the young man, who had watched him chasing a butterfly, took him by the hand, led him to the hut and took down an object transfixed to a post. He displayed it with an air of great mystery; it was a large chrysalis suspended from a leaf that he placed carefully in Bates hands, saying, *Panapana curi* (Tupi: butterfly by-and-by). Fascinated, Bates realised these people knew the metamorphoses of butterflies, but being unable to talk in his new friend's language, he could not establish the full implications of this fact.

He moved on once more and eventually the schooner reached Fonte Boa, where Bates went ashore and stayed for nearly two months until the next steamer stopped on its way downriver. Fonte Boa was like most other settlements on the upper Amazons; it lay on a secondary waterway, this time a narrow stream as straight as an artificial canal called the Cayhiar-hy. It was no more prepossessing than Tonantins, being just another wretched, muddy, dilapidated village, as Bates told the readers of *The Zoologist* a few months later. It was built on a clay plateau, and arranged about a square that had been neglected and allowed to revert to bush, so that the whole place was thickly carpeted with a tough growth of shrubs. During the rains, it was an impassible glutinous swamp; in the dry season nothing more than a dust bowl. Bates found somewhere to stay, a house much in keeping with this awful place, the plaster walls were lined with a crop of green mould, and slimy moisture oozed through the black, dirty floor; the rooms were large, but lighted by miserable little holes in place of windows. Outside the house was a solid path two feet wide, but 'to step over the boundary, formed by a line of slippery stems of palms, was to sink up to the knees in a sticky swamp'. Bates thought the people were rather like the place, run down and dilapidated, a 'loose-living, rustic, plain-spoken and ignorant set of people.'[110]

There was no priest, schoolmaster, or obvious government in Fonte Boa. The leading trader, who had built an ostentatious house with brick floors and a tiled roof, was so ignorant that he snipped the illustrations out of a bundle of *Illustrated London News* Bates had lent him, and pasted them, many upside down, on the newly white-washed walls of his house. When Bates, accepting a fait accompli, allowed him to keep them, he was so delighted he presented Bates with a whole boatload of live turtles.

Mosquitoes overran the damp rooms and Bates was constantly bitten during the daytime through his thick trousers, but the great calico tent provided fair protection at night. In the forest, a larger and more forceful species followed Bates in a little cloud, humming so deafeningly that he could not clearly hear the birds around him. My experience was similar to Bates's; however, despite the mosquitoes and the dreaded flies Bates still seemed to enjoy himself.

Although Bates was tired and had been concerned about his health for a long time at this moment he was robust. The persistent mosquitoes were manifestly not malaria carriers and the forest was extraordinarily beautiful,

[110] Bates, H W, 'Proceedings of natural history collectors in foreign countries, Mr. H W Bates'. *The Zoologist*, 11, (1853), 3726-3729.

with lush vegetation and the most gigantic trees Bates had ever seen. 'Through the shades', he reported to *The Zoologist*, 'flowed two sparkling brooks, with crystal waters, a feature rare to meet with on the Upper Amazons, where the small water-courses generally dry up in the dry season.' In addition to many interesting birds and mammals, including the Umbrella Chatterer and the Curl crested Toucan and a remarkable bear-like monkey, *Pithecia hirsuta*; he collected no less than a hundred hitherto unknown species of insects at Fonte Boa.

On the 25th January 1857, the steamer arrived and Bates boarded. After sixteen hours he was back in Tefé, thankful to return to his favourite place on the Amazons, with its white sandy shores, cropped grass, neatness and cleanliness. It was the dry season and he was to enjoy almost nine months collecting there until 5th September 1857, when ever eager to explore the farthest regions of the river westwards, he set sail once more on the *Tabatinga*, this time on route for São Paulo de Olivença, four hundred miles west of Tefé. At the furthest extremes of the river and getting closer to its source, Bates found the side channels dry with deep white sandy ravines where streams had been before in the midst of the thick lush forest. On the sand were ducks, storks, herons, spoonbills, and other aquatic birds, while alligators congregated in great numbers among them.

Hauxwell was on board which pleased Bates and meant that for a while he had another intelligent conversationalist travelling with him, a luxury he was determined to exploit to the full. Bates amused himself taking pot shots at the birds with Hauxwell's double-barreled shotgun scoring an amazing number of effective hits. They also caught a huge caiman which:

> "...the commandante resolved to haul on board for the purpose of extracting a part of the animal, which is in great repute among Brazilians as a 'remedio'; for this purpose he stopped the steamer and sent a boat; the men in the boat had some difficulty in towing the beast, and it took eight or ten strong men to get it on deck. It had still some remains of life, and caused great commotion on board when it lashed its heavy tail and opened its ponderous red jaws; a blow with a hatchet on the crown easily composed him at last. The length was fifteen feet, but this cannot give a correct idea of the immense bulk of the animal, as the head and trunk are much larger in proportion than they are in the smaller animals of the lizard tribes generally."

Of all the destinations to the west of Tefé, São Paulo de Olivença was the largest Bates visited. Only 500 inhabitants lived there and it turned out to be another run down settlement with few graces. The streets were narrow, dirty, and dusty, having been deep in mud through the wet season and the houses were ruinous. The noises of the forest, made mostly by frogs and toads in the weedy neglected gardens, set up such a chorus after sunset that even indoors conversation had to be carried on by shouting. The house Bates found was even worse than that in Fonte Boa. It was wet inside rather than damp, and the air so humid he had the greatest difficulty preventing his specimens from turning mouldy. What shocked him however was the life-style of the inhabitants who seemed to indulge in a perpetual orgy and prompted his comments that he 'never saw anything so disgusting in the course of my travels.'

The inhabitants were a mixture of poor mamelucos and degenerate Indians of the Tacúna and Gollina tribes, all of whom seemed to be in a state of perpetual drunkenness, their slovenly habits making the place even worse than it might otherwise have been. Two Negroes who were levelheaded in their outlook became his regular companions, one the sub-delegado of police Jose Patricio, the other a tailor named Mestre Chico whom he had known in Belém. Bates found the chief residents of the town just as profligate. The priest gambled the nights away, while the Justice of the Peace, Geraldo, and the Director of Indians, Ribeiro, were inveterate drunkards.

> "Geraldo and Ribeiro were my close neighbours, but they took offence at me after the first few days, because I would not join them in their drinking bouts, which took place about every third day. They used to begin early in the morning with Cashaua mixed with grated ginger, a powerful drink, which used to excite them almost to madness. Neighbour Geraldo, after these morning potations, used to station himself opposite my house and rave about foreigners, gesticulating in a threatening manner towards me by the hour. After becoming sober in the evening, he usually came to offer me the humblest apologies, driven to it, I believe by his wife, he himself being quite unconscious of this breach of good manners. The wives of the São Paulo de Olivença worthies, however, were generally as bad as their husbands; nearly all the women being hard drinkers, and corrupt to the last degree. Wife beating naturally flourished under such a state of things. I found it always best to lock myself in-doors

> after sunset, and take no notice of the thumps and screams which used to rouse the village in different quarters throughout the night, especially at festival time."

Bates stayed there five months, finding the natural abundance of species at São Paulo de Olivença as prolific as its vices. Ever the professional naturalist he thought five years would not have been long enough to exhaust the treasures it offered the natural history collector.

He had now been ten years in the South American forests, but the scenery here was outstanding giving him 'as much enjoyment as if I had only just landed for the first time in a tropical country'. The forests were dense but interlaced by broad forest rides and clear, cool brooks - ideal places to attract brilliant birds and butterflies. 'I had the almost daily habit, in my solitary walks, to resting on the clean banks of these swift flowing streams, and bathing for an hour at the time in their bracing waters; hours which now remain amongst my most pleasant memories.'

Bates collected few birds, since he could not find Indian hunters willing to work for him, but the number and variety of insects he collected were extraordinary. Another 5,000 superb specimens, including no less than 686 new species, of which 79 butterflies hitherto unknown to science were added to his collection. Among the butterflies, he found surprising variation in the species from Fonte Boa. He paid special attention to the genus *Ithomia* and, capturing nineteen species, found that eight of them were new to him, whereas eight other species of the same genus which he had newly discovered at Fonte Boa were all absent in São Paulo de Olivença. These butterflies led Bates to his most interesting observations on mimicry that he published in *The Zoologist* early in 1858, revealing the ideas he was formulating more than a year before the appearance of Darwin's *On the Origin of Species* and three years before he published his own theory of mimicry:

> "Flying amongst the Ithomiae was now and then to be observed a Leptalis; I was very careful to secure every specimen, and the gathered series, now I come to examine them closely, have interested me as much as any other acquisition made during my excursion. [Apart from] A white species, the rest may be considered either as six species allied to L[eptalis] lysinoe [Hezeits] or as the latter branching out into six rather widely differing varieties. In either case they are very interesting, because some of the kinds come to imitate, each a

species of Ithomiae common only in this locality...It would seem then almost correct to say, that at Ega and other stations these new Leptales are not found, because the Ithomiae to which they correspond are also absent. L. lysinoe imitates Ithomia flora; but three at least of the new species imitate three of the commonest Ithomiae of St Paulo; on the wing their resemblance is much more striking than when in the cabinet. In fact I was quite unable to distinguish them on the wing; and always on capturing what I took for an Ithomia, and found when in the net to be a Leptalis mimicking it; I could scarcely restrain an exclamation of surprise...The resemblance between Leptales and Ithomiae, two groups of Diurnes much more widely separated than they appear in our classifications, is repeated in the case of a group of Bombycidae moths, of which there are at least two genera imitating the Ithomiae and the larger Heliconiae. One of them, which I saw first at the British Museum, exactly imitates Ithomia flora; at Ega, there is one imitating in the same way Ithomia fluonia of the same locality...These analogies to me appear one of the most beautiful phenomena in Nature."[111]

Bates was an absolute master at describing complex issues in language that could be understood by the non-professional and was a pioneer in the art of popularising science. However, he expressed himself with care for the specialist reader, presenting facts he had observed without going on to supposed conclusions. Nevertheless there were always enough clues in his writing to show the general direction of his thoughts. The curious relationship between varieties and species, sometimes hard to distinguish even by the expert, is clearly in his mind, and though the word *adaptation* is never used when he discusses the phenomenon of imitation, the idea is evident.

In an interesting letter which Bates wrote to Joseph Hooker three years after he left the Amazons (5th March 1862), he makes it clear that he already had accepted the fact of evolution at this time, explaining it not by natural selection, but by the shaping influence of the environment:

"While I was travelling I used to attribute, like many others, the production of distinct local varieties, or races, to the direct action of

[111] Bates, H W, Contributions to an insect fauna of the Amazon valley. Lepidoptera-Papilionidae, *Journal of Entomology* (1861), 1, 218-245.

the local conditions; and when I found such coming in contact with their supposed parents without intermarrying or reverting, I supposed the cause was the gradual and slow change in their constitutions which had brought them to a point when they were incapable of going back." [112]

Close to the settlement where he now was, more primitive Indians, the Tucunas, inhabited the forests, many of whom still lived in the traditional manner of their people, almost untouched by the world around them. The whole community lived in a single hut called a malocas. It was oblong:

"...built and arranged inside with such a disregard of all symmetry that it appeared as though constructed with a number of hands, each working independently, stretching a rafter or fitting in a piece of thatch, without reference to what his fellow-labourers were doing. The walls and roofs of these houses were thatched with palm leaves; hammocks were slung between posts that supported the roof, and overhead were stages of split palm stem where, on the onset of puberty, the girls were sent to sit in the smoke until they were thought to be purified; alas some died under the treatment."

Initially the Indians were shy, running away from Bates into the forest whenever contact seemed to be getting too close. However Bates simply went about the business of collecting his butterflies and, before long, curiosity made the Indians throw caution to the winds and approach him. In time Bates found them to be trusting, friendly and good-natured. He discovered that they were not completely primitive. They had some practical skills. For example, they were expert potters making large painted jars. Some jars were used for food storage. They stuffed their dead chiefs into others for burial under the floors of their houses.

Feasts and ceremonies which were held either to pay tribute to Jurupari, their principal deity, or to celebrate such occasions as weddings, the feast of fruits, or the ritual plucking of hair from children's heads, played an central role in their culture. They wore no clothes, only bracelets. Anklets and garters of tapir-hide or tough bark adorned their bodies but on feast days they would dress themselves in elaborate ceremonial dress:

---

[112] Letter to Joseph Hooker, 5th March 1862, Royal Botanical Gardens Archives, Kew.

"The chief wears a headdress or cap made by fixing the breast-feathers of the Toucan on a web of Bromelia twine, with erect tail plumes of macaws rising from the crown. The cinctures of the arms and legs are also then ornamented with bunches of feathers. Others wear masked dresses: these are long cloaks reaching below the knee, and made of thick whitish-coloured inner bark of a tree, the fibres of which are interlaced in so regular a manner that the material looks like artificial cloth. The cloak covers the head; two holes are cut out for the eyes, a large round piece of the cloth stretched on a rim of flexible wood is stitched on each side to represent ears, and the features are painted in exaggerated style with yellow, red and black streaks. The dresses are sewn into the proper shapes with thread made of the inner bark of the Uaissima tree. Sometimes grotesque headdresses, representing monkeys' busts or heads of other animals, are worn on these holidays. The biggest and ugliest mask represents the Jurupari. In these festival habiliments the Tucunas go through their monotonous see-saw and stamping dances accompanied by singing and drumming, and keep up the sport often for three or four days and nights in succession, drinking enormous quantities of caysuma, smoking tobacco and snuffing parica powder."

It was provident that, towards the end of his travels in the Amazons, Bates met these appealing people. This was the closest he was ever to get to the primitive as hostility usually prevented such contact. He knew other fêted tribes, such as the fierce Araras of the Madeira, only by repute or through having encountered individual members, as in the case of the Majeronas, whose hostility to travellers prevented him from following the course of the Jauari beyond São Paulo de Olivença.

A story was being recounted at this time of two half-breed travellers who ascended the river into the country of the Majeronas not long before Bates arrived in São Paulo de Olivença, when they made advances to the women of the tribe they had been shot down with arrows, roasted and eaten. Bates's friend, José Patricio, organized a retaliatory expedition, but found the tribe departed, except for one girl they brought back with them. As was the custom with captured tribespeople she was taken to São Paulo de Olivença, baptized and integrated into a household as a servant, on this occasion into José Patricio's home. This compassionate approach is rather similar to the way the Brazilians treated their slaves before abolition. Since Bates was now without his own servant, she was sent daily to his house to fill the water pots and make the fire. This allowed him to study the girl who was from a cannibalistic tribe. He won her friendship by extracting a gadfly grub from her back that had raised a painful tumour:

> "I heard this artless maiden relate, in the coolest manner possible, how she ate a portion of the bodies of the young men whom her tribe had roasted. But what increased greatly the incongruity of this business, the young widow of one of the victims, a neighbour of mine, happened to be present during the narrative, and showed her interest in it by laughing at the broken Portuguese in which the girl related the horrible story."

Mind over Matter

To Bates, these forests of the upper Amazons were part utopian retreat from the onslaught of reality, a realm of peace and beauty, and part a symbol of isolation, solitude, and even death. Beyond being tangible, the forest was a lonely place and Bates must have experienced desperate longings for company and home. Like Marlow in Conrad's *Heart of Darkness*,[113] Bates encounters darkness in São Paulo de Olivença that is both moral and physical. He was exhausted by the journey that had started back in 1848 and by the continuous deprivations and sickness that plagued him. The place, so beautiful in its scenery, so abundant in its wild life, and so near the insouciant innocence of primeval man, was to be the grave of Bates final plans, and almost of Bates himself. What he wanted more than anything else was to travel on up river another six hundred miles to the foot of the Andes at Moyabamba which would have completed his original plans for examining the natural history of the Amazons to its source. He also wanted to add to his collection specimens from yet another little known area. Now, in his fourth month in São Paulo de Olivença, he became extremely sick with a particularly virulent attack of malaria, so ill in fact that those around him considered he would die. The strain of malaria endemic in the Amazon in the 1850s was *falciparum*, brought to Brazil by West African slaves, and more deadly than the *vivax* form of the northern hemispheres. He dosed himself

[113]Conrad, Joseph. *Heart of Darkness*. (London: Wordsworth Classics, 1994).
In this story, Conrad presents Marlow as a traditional hero: tough, honest, an independent thinker, a capable man. Yet he is horrified by the debauchery he discovers on his journey up the river Congo and becomes exhausted, disbelieving, and contemptuous of those he encounters. Conrad also uses Wallace as his model for Marlow in *Lord Jim*.

with quinine and chamomile tea but his body, weakened by bad food, overwork and the harsh climate, could barely resist the illness. As soon as he was able, he left his bed, frightened of the possibility of an inflamed liver that could have meant death, and with great courage, despite his weakness, would gather his collecting equipment and his gun and walk into the forest each morning in search for more specimens. He was seized by tremors and shivering each day but just stood still to brave it out before continuing his pursuit.

Lieutenant Nunes saved Bates, for on arriving with his boat the *Tabatinga* in January he was horrified to see the state of his new friend. Bates talked to Nunes about his plan to stay at São Paulo de Olivença.in the hope of recovering sufficiently to continue his journey westward, and reaching the elusive objective of his expedition. Nunes however convinced Bates that the only real hope of recovery would be to return to Belém and eventually England by the quickest possible means.

When the steamer sailed down river in February 1858, Bates took Nunes' advice and went back to Tefé. There the fever disappeared, leaving a weakness that made all activity seem burdensome. Bates stayed on in Tefé, month after month, unwilling to abandon his plan to 'gather the yet unseen treasures of the marvellous countries lying between Tabatinga and the slopes of the Andes'. He had already explored the main river and its side waters for 1,800 miles from Belém to São Paulo de Olivença and was reluctant to leave with the task unfinished. However, his health never recovered sufficiently for the trip to Moyabamba, and after a further year in Tefé, he decided he had no real alternative but to return home.

On 3rd February 1859, he bade farewell to his friends in Tefé and, accompanied by José and Sebastian and a tame scarlet-faced Uakari monkey, boarded a schooner that was making the journey downriver. On 17th March, Bates arrived at Belém where he was the guest of G. R. Brocklehurst, now the leading foreign merchant. From his host's house, during the ten weeks he remained in Belém, he re-explored the city which, during the 1850s, had prospered and improved almost beyond recognition. There were bookshops and libraries, newspapers and cabs. Squares and streets had been paved, and most of the old dilapidated buildings replaced. Belém had caught up with the century.

Both he and the city had changed profoundly since he had been there seven and a half years before. His old friends, who had seen him depart as a fresh-faced young man, hardly recognized the 'oldish, yellow-faced man with big whiskers' as he described himself in a letter of gentle warning to his parents. However, he was made much of, the interior of the country was still to the people of Belém a terrifying wilderness, and 'a man who has spent seven and a half years exploring it solely with scientific aims was somewhat of a curiosity'. As for the city itself, much that had attracted Bates when he first set eyes on Belém on that brilliant May dawn of 1848 had been swept away. The religious festivals were less barbarically splendid, and so were the people. Prices had risen immoderately and only the rich could afford servants. Worst of all, in Bates's eyes, was the clearing of vegetation from the suburban lanes and the driving of carriage roads through the forest, so that the old overgrown paths where he and Wallace had once wandered were no more and their wild inhabitants had vanished.

In following Bates on this long voyage through not only the rainforests of Amazonia but also the evolution of the man, we have to ask certain questions. There is some spin in the manner in which the book, *The Naturalist on the River Amazons*, deals with aspects of the story, no doubt because Murray the publisher would have known better than anyone else what would sell the work. Nevertheless, Bates as a Victorian was writing in an era when sensibilities above all else would have determined his style.

For example who was Oria, the girl child Bates was so compassionate and emotional about? Is there any hidden meaning when Bates writes about the old Indian and 'ecui encommenda? Was Bates, after eleven and a half years in the wilderness, simply sentimental about the Indians to the extent that he acted in this extraordinary way out of kindness or was this his own child? Would he have been so emotional about a child he had only known for a few weeks?

However according to Professor Moon there is no foundation for supposing that Bates acquired poor standards of behaviour on the Amazons, which could subsequently be associated with the birth of an illegitimate child... In an article in *The Adelphi,* Woodcock writes:[114]

> "... of his sexual life, unfortunately, we gather nothing, except for a few diffident references to the attractiveness of certain of the women he encountered."[115]

---

[114]Moon, H P, *Henry Walter Bates FRS, 1825-1892: Explorer, Scientist and Darwinian.* (Leicester: Leicester Museums, Art Galleries and Records Service. 1976) 48.

[113]Woodcock, G, Henry Bates on the Amazons. *Adelphi* (1945) 115-121.

Miscegenation perhaps?

Nevertheless, quite natural behaviour could have resulted in such happenings. Travellers like Bates usually met Indian men on the riverbank or deep in the forest whilst the former were going about their business of hunting, shooting with the bow and arrow and fishing. Often the Indians would run away at the sight of a white man but the reaction of the women among them was quite often different. Bates always commented on the women but not as liberally as some of his contemporaries. Women were constantly busy about the village, cooking, spinning and weaving, raising families and tending crops. Bates was clearly as struck by the beauty of the women as was Carl von Martius (1794-1868) who admired the Pase women of the Japura River:

> "The wife of Chief Albano had such regular features, such brilliant black eyes, and such a perfectly formed body that, with her blue-black tattooed lower face, she would have caused a sensation even in Europe."

According to John Hemming 'Farther upstream, Martius noticed the beautiful figures of the Aruak speaking Yumana women - indeed, it was for their figures that the Yumana and nearby Marawa were eagerly sought as slaves by settlers.'

Martius observed that Indian women wore clothes only when strangers were present.

Even when the missionaries had influence over these women, they only covered themselves when in the company of missionaries or strangers and on the departure of these cast off all their modesty. Bates recalled a time he particularly enjoyed when he was with the Mundurucus and surrounded by a group of thirty or more such naked beauties who were entranced by the pictures in a book he was showing to them.[116]

According to the Italian traveller Ermanno Stradelli (1852-1926) this practice had not changed by the turn of the twentieth century 'their clothing is nothing more than an ornament, of which they are proud in front of those who teach them to wear it, but which they abandon once the whites have passed.' When Stradelli witnessed this, he observed that the Tukano (Tukana) women of the Uaupes wore only a minuscule tanga-fringed girdle

[116] Stradelli, Count Ermanno, O rio Branco, O rio Branco. Cathedral Archives, Tefé.

when dancing, but were otherwise completely naked, 'with their labia major tucked in and the mons veneris completely devoid of hair, which they pluck with pincers of split pieces of uambe liana.'[117]

Travellers generally found that acculturated Indian women could be equally delightful. The English explorer William Chandless (1829-1896) admired the agreeable behavior of the Munducuru women on the Jutai River. 'Their brightness and vivacity were unconscious and quite distinct from forwardness. It seemed to me to be the elasticity in the first freedom from woman's bondage of tribe-life, and probably may pass away; for those on the Maue-assti had nothing of it: indeed they were dull and shy'. According to Hemming:

> "Araujo e Amazonas said that civilised Indian women held their long black hair with semicircular high-fronted gold combs. A white kerchief was worn on either the hand or the shoulder. These women used perfumes from forest plants and wore small bouquets of flowers. "The lady's good taste and coquetry is shown by where she places this garland on her head, at one side or as a pendant." Amazon Indian women were modest and pious, observing all church festivals and dancing less voluptuously and more monotonously than in other parts of Brazil."[118]

However, what Bates noticed in his early days in Belém was the acculturated women's blouses, a feature that most excited onlookers. The sleeves were wide and buttoned at the wrist with colourful or gold buttons, which caused the sleeve to fall over the wrist. The blouse was loose and usually made of diaphanous material that left little to the imagination of the onlooker who could appreciate in full the wearer's attributes. 'They were made of lacework or muslin so fine that it shades rather than covers the breast.' Another later traveller, Robert Ave-Lallement, (1812-1884), could not take his eyes off these bright transparent shirts. 'At every step, the fine cloth of the blouses buttoned at the neck quivered over firm and elastic breasts, whose exuberance needed no corset to support them.'

The Mameluco women observed by Araújo e Amazonas were even more enticing. He found them as attractive in their manner and culture as in their physical beauty, which suggests that the longer the traveller stayed in the

[117] Royal Geographical Society Minute Book 1866, Ascent of the River Purus.
[118] Hemming, John, *Red Gold, The Conquest of the Brazilian Indians*, (London: Pan Books, 2004).

region the more acceptable the simple uneducated women became. There is no question in my mind that the younger women were physically beautiful and all possessed a wild manner that itself was alluring:

> "Small hands and feet, ample black tresses, the proverbial necks of Indian women, figures of the most perfect proportions, added to a tanned colour: all these were enhanced by an original facial expression, and by vivacity and grace infinitely superior to what one would have expected for a race isolated in that wilderness."[119]

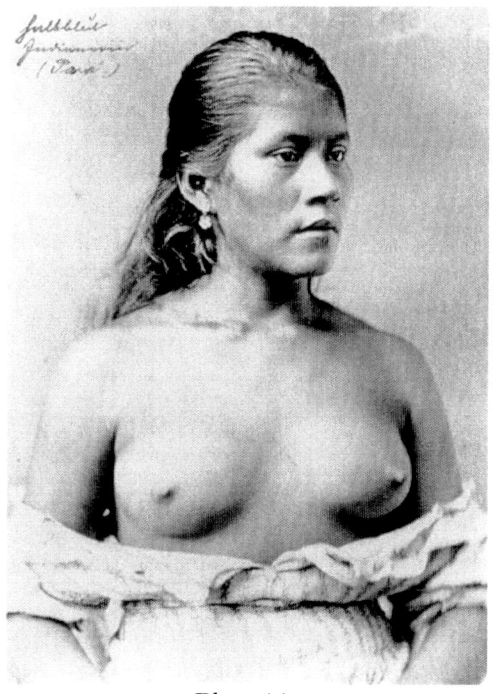

Plate 11.
An Indian woman in Para by George Huebner c 1890

Most Indian women wore some ornament even if it only came from the forest such as a flower in their hair. They also obtained ornaments and trinkets from itinerant traders, as enterprising salesmen were the main means of contact between *civilizations* and isolated riverside settlements. The roving

[119] Academy of American Franciscan History, *Aragon e Amazonas*, Cathedral Archives, Tefé.

merchants, known as regatao, mercilessly drove a hard deal and many were unscrupulous rascals into the bargain. They travelled in fleets of montarias, riverboats paddled by Indians and mestizos. Their crews, called canicurus, would often get drunk and behave disgracefully towards the women, which more often than not led to a quarrel. The regatao traders with impunity perpetrated many indignities and crimes on their gullible customers. William Edwards with whom Bates and Wallace talked before they departed for the Amazon commented that the regatao traders on the Amazon in 1846, sold:

> "...coarse German and English dry goods, Lowell shirtings, a few descriptions of hardware, Salem soap, beads, needles, and a few other fancy articles ... In return are brought down balsam, gums, wax, drugs, turtle-oil, tobacco, fish and hammocks."[120]

Even young Herbert Wallace, whose obsession was poetry, put his thoughts and experiences into verse. 'Lasses darkly delicate/with eyes that ever kill/breathed to him 'in whispers/that we are in Brazil.' How could Bates not notice the women that he encountered on the Amazons? In the absence of any evidence to the contrary we can safely assume that as a full-blooded, normal, young, healthy, heterosexual male in his prime, he would have more than noticed them.[121]

The recent appearance of octogenarian Eskimo sons of the polar explorers Robert Peary and Matthew Henson suggests that explorers in the field were not as celibate as has generally been supposed. With an Inuit woman named Akatingwah, Matthew Henson fathered his only child, a son named Anaukaq. After 1909, Matthew Henson never saw Akatingwah or his son again though he did receive updates about them from other explorers for a time.[122]

Wallace on the River Uaupes was not the first scientific traveller to arrive from Europe. At the last house on route to the Serra he saw a young mameluca, 'very fair and handsome, and of a particularly intelligent expression and countenance'. Evidently, she was the daughter of the celebrated German naturalist Dr Natterer, (1787-1843): 'She was,' said

---

[120] Edwards, William H, *A Voyage up the River Amazon, Including a Residence at Pará*, (London: John Murray, 1847).
[121] Hemming, John, *Red Gold, The Conquest of the Brazilian Indians*, (London: Pan Books, 2004) 44.
[122] The discovery of Anauakaq and Kali was documented in a book entitled *North Pole Legacy: Black, White, and Eskimo.* AS Allen Counter.

Wallace, 'a fine specimen of the noble race produced by the mixture of the Saxon and Indian blood.'[123]

Travellers often saw miscegenation with local women as convenient and pleasurable; there is no reason to suppose Bates would have felt differently. Sexuality is recognised as a prime mover among explorers' unconscious drives. Because of the isolated areas where Bates was exploring on the Amazon, and bearing in mind Victorian prudery, it is not easy to establish the truth. However oral tradition and common sense endorse the conclusion that liaisons with native women could have happened. Indeed, one of the motives for living on the frontier was to escape from the trammels of civilisation, its discontents and prudery, and the greatest liberty of all was sexual freedom and indulgence. In order to get inside a culture easily and learn about its idiosyncrasies, unspoken codes, and deep structures, or just for companionship and because of utter loneliness, such liaisons were commonplace amongst travellers. The young women Bates encountered were without doubt attractive and as time passed would have become more so to the lonely and sometimes inescapably sexually frustrated traveller. Indian women were ready partners, and in the early days (of exploration), their men acquiesced because it was hospitable to offer women to strangers.

The great Jesuit missionary João de Anchieta wrote to his Superior that:

> "...the women go naked and are unable to say no to anyone. They themselves in fact provoke and importune the men, sleeping with them in hammocks; for they hold it to be an honour to sleep with the Christians.' Another shocked Jesuit reported that the women: 'have little resistance against those who assault them. In fact, instead of resisting, they go and seek them out in their houses! The women go naked and are so base that they go after the youths to sin with them and seduce them. And these readily allow themselves to be seduced."[124]

In the midst of such easy sexuality, it would be surprising if Bates, despite the inhibitions and moral scruples about promiscuous intercourse resulting from both culture and upbringing, had not had some intimate relations with the opposite sex, during his eleven years in Brazil.

---

[123] Wallace, Alfred Russel, *Travels on the river Amazon and the Rio Negro*, (London: Ward Lock & Co, 1889).

[124] Hemming, John, *Red Gold, The Conquest of the Brazilian Indians*, (London: Pan Books, 2004) 44.

The morality issue was that women were crucial to the maintenance of the boundaries between races, by engaging or not engaging in sexual activity. Equally, they were central to controlling transgressions of these boundaries through an ungoverned sexuality. Since the tropics were perceived as a place where the constraints of civilisation were loosened, and the heat of the climate created an unbridled sensuality, to the traveller a naturally naked tropical woman was a symbol of sexual accessibility and a liaison, although it might have threatened the sense of racial order, could have been irresistible.

A clear account of Amazon Indian attitudes to such liaisons is given in Philippe Descola's book *The Spears of Twilight, Life and death in the Amazon Jungle.* Living close to and sometimes with the Indians as Bates had done, Descola explains the attitude of the genders to each other and the morals that prevailed in their societies. Love for the Indian he explains is fuelled in a mutual attraction to the body of one's companion and the physical qualities that make it desirable, but it is also the fruit of an attachment to moral and social virtues that are quite precisely defined. In a man, women appreciated eloquence, courage, energy, a happy temperament, musical talent, excellence in hunting, and technical skill, all attributes that could easily have applied to Bates. We can therefore assume the women he met would have looked at him favourably. Desirable qualities in a woman, were a smiling modesty, biddableness, gentleness, skill at domestic tasks, and being a good gardener, weaver, and potter, domestic qualities that would have appealed to Bates as indeed they did when, much later, he chose Sarah as his bride.[125]

Descola goes on to say that in the first place it was usual for men to take wives from the children of their maternal uncles or their paternal aunts. Furthermore, families were so large that virtually everybody was bound to have at least a dozen potential consorts in this category. Male and female cousins played together as children and developed friendships that, during adolescence, led to more long-lasting feelings of affection or antipathy. When boys told their father, or girls told their mother, about their feelings towards one of their cousins, it probably led without complications to co-habitation.

Beyond this, the circle of relationships available to the young, whilst plentiful, was inward looking. Given the extremely dispersed habitat of the

---

[125] Descola, Philippe, *The Spears of Twilight, Life and Death in the Amazon Jungle.* (London: Harper Collins 1993) 182-186, 260-262.

forest, the hostility between many neighbouring groups and the resultant insecurity, everything conspired to limit choice. Inbreeding was therefore a difficulty and any opportunity to widen selection was enthusiastically accepted as a gift of fate.

Descola makes the point that his Indian acquaintances were always rather sorry that he only had one wife and several men had given him to understand they would happily give him one of their daughters for sex.

To the native Indian, celibacy was meaningless, even an object of mocking commiseration when, as in the case of the missionaries for example, it was attributed to physical infirmity or impotency. When it originated through deliberate solitude as might have been the case with Bates, it could equally arouse suspicion. However the native Indian did not favour casual sexual relations. An Italian colleague of Descola's told him of the distrust he encountered whenever he arrived in an Indian's house and the extreme precautions he had to take when speaking to a woman, in order to avoid being suspected of casual seductive designs. Much better a wife should accompany him even if she was a native girl. A perceived marital relationship gave the Indian a chance to satisfy curiosity regarding the traveller's customs simply by observing, and to some extent this tempered the general anxiety felt about foreigners. As the Indians said to Descola:

> "Why do the Whites say that you should not have several wives? They have mistresses and they pay to sleep with tarts, whereas we do not change wives, we simply add to them...It is very hard having only one wife. There are times when you cannot make love to her, when she has just had a child and is not yet 'dry', when she has the 'moon blood' [period] or when she is ill. With several wives, we no longer have to suffer."

What better circumstances could there have been for the traveller to conform to native practices and take a native girl with the approval of the Indians.

Within two years of Bates returning to civilisation he had made a local butcher's uneducated 17 year old daughter pregnant. At this time, a single girl who became pregnant brought disgrace to her family who would have insisted, if possible, that the father married the girl. Bates eventually did so but his family disapproved of the match. She was they said, 'not from his class or walk in life'. However the subsequent marriage, which did not take place until almost a year after their child's birth, lasted until Bates's death and appeared to be successful. It is pure speculation but normal conduct would suggest that Bates must have had some sexual relationship whilst in Brazil. The uncomplicated openness of the simple innocent girls of the Amazons, suggests that the story of Oria could have had such beginnings. Perhaps Bates even wanted his Victorian readers to think about him in this way.

In chapter XIII of *The Naturalist on the River Amazons* there is an artist's representation based on Bates's account of the masked dance of the Tacúna Indians at a wedding festival in Tefé where he was closely involved with the tribe. Bates took a deal of trouble over this woodcut by J B Zwecker, (signed on the left hand side) as with all the plates in his book, even asking Darwin for his opinion.

Plate 12.

Tacúna Indians at a wedding festival in Ega

At one stage, Darwin considered paying £10 towards the cost of the plates. It is safe to assume therefore that Bates was satisfied with the results. It is of course a sketch and not a photograph and therefore it is contrived. In this plate, Bates is in the background on the left hand side wearing a check shirt. Admittedly, he is stockier in this drawing than in real life but having approved the illustration, he must have fully understood its implications. Here we see him at ease enjoying a drink close to a completely naked young girl. The only person observing her is another woman in a hammock above. All the other men and women are fully occupied with the event and looking to the right or out of the picture. None of them is observing Bates. As Descola says in his narratives, circumstances like this would not have been tolerated unless the men felt confident that their women were not at risk. Their sole criteria for this would have been that Bates had a woman of his own, most probably suggested by the girl in the sketch, and so it would be reasonable to assume that they knew he would not covet their women.

Chapter 7

The Incredible Journey was over

Bates packed up the rest of his collection and shipped it to England, dividing it into three consignments on three different ships so it would not suffer the same fate as Wallace's. He then arranged a passage to England for himself via New York on an American boat, the *Frederick Demming*.

His own words are exceptional in recounting the twilight hours he spent in Belém and on the Amazons:

> "In rambling over my old ground in the forests of the neighbourhood, I found great changes had taken place - to me, changes for the worse. The mantle of shrubs, bushes, and creeping plants which formerly, when the suburbs were undisturbed by axe or spade, had been left free to arrange itself in rich, full, and smooth sheets and masses over the forest borders, had been nearly all cut away, and troops of labourers were still employed cutting ugly muddy roads for carts and cattle, through the once clean and lonely woods. Houses and mills had been erected on the borders of these new roads. The noble forest trees had been cut down, and their naked half-burnt stems remained in the midst of ashes, muddy puddles, and heaps of broken branches. I was obliged to hire a Negro boy to show me the way to my favourite path near Una, which I have described in the second chapter of (t)his narrative; the new clearings having quite obliterated the old forest roads. Only a few acres of the glorious forest near Una now remained in their natural state. On the other side of the city, near to the old road to the rice mills, several scores of woodsmen were employed, under government, in cutting a broad carriage-road through the forest to Maranham, the capital of the neighbouring province, distant 250 miles from Pará, and this had entirely destroyed the solitude of the grand old forest path. In the course of a few years, however, a new growth of creepers will cover the naked tree-trunks on the borders of this new road, and luxuriant shrubs form a green fringe to the path: it will then become as beautiful a woodland road as the old one was. A naturalist will have, henceforward, to go farther from the city to find the glorious forest scenery which lay so near in 1848, and work much more laboriously than was formerly needed, to make the large

collections which Mr. Wallace and I succeeded in doing in the neighbourhood of Pará."

Eventually on 2nd June 1859, Bates left Belém for New York, the North American route being a fast, safe, and most pleasant way of reaching England. The next evening he took a last view of the glorious forest for which he cared so much, and to which he had devoted so many years. There was no turning back:

> "The saddest hours I ever recollect to have spent were those of the succeeding night, when, the mameluco pilot having left us free of the shoals and out of sight of land, though within the mouth of the river, at anchor, waiting for the wind, I felt that the last link which connected me with the land of so many pleasing recollections was broken. The Paraenses, who are fully aware of the attractiveness of their country, have an alliterative proverb, 'Quem vai pará (o) Pará Pará,' 'He who goes to Pará stops there,' and I had often thought I should myself have been added to the list of examples. The desire, however, of seeing again my parents and enjoying once more the rich pleasures of intellectual society, had succeeded in overcoming the attractions of a region which may be fittingly called a Naturalist's Paradise. During this last night on the Pará River a crowd of unusual thoughts occupied my mind. Recollections of English climate, scenery, and modes of life carne to me with vividness I had never before experienced during the eleven years of my absence. Pictures of startling clearness rose up of the gloomy winters, the long grey twilights, murky atmosphere, elongated shadows, chilly springs, and sloppy summers; of factory chimneys and crowds of grimy operatives, rung to work in early morning by factory bells; of union workhouses, confined rooms, artificial cares, and slavish conventionalities. To live again amidst these dull scenes I was quitting a country of perpetual summer, where my life had been spent like that of three-fourths of the people in gipsy fashion, on the endless streams or in the boundless forests. I was leaving the equator, where the well-balanced forces of Nature maintained a land-surface and climate that seemed to be typical of mundane order and beauty, to sail towards the North Pole, where lay my home under crepuscular skies somewhere about fifty-two degrees of latitude. It was natural to feel a little dismayed at the prospect of so great a change; but now,

after three years of renewed experience of England, I find how incomparably superior is civilised life, where feelings, tastes, and intellect find abundant nourishment, to the spiritual sterility of half-savage existence, even if it were passed in the garden of Eden. What has struck me powerfully is the immeasurably greater diversity and interest of human character and social conditions in a single civilised nation, than in equatorial South America, where three distinct races of man live together. The superiority of the bleak north to tropical regions, however, is only in its social aspect; for I hold to the opinion that, although humanity can reach an advanced state of culture only by battling with the inclemencies of nature in high latitudes, it is under the equator alone that the perfect race of the future will attain to complete fruition of man's beautiful heritage, the earth."

The following day there was little or no wind and the *Demming* drifted in the current of fresh water that pours from the mouth of the river. In twenty-four hours the ship advanced only seventy miles. On 6th June when about 400 miles from the mouth of the main river, it passed numerous patches of floating grass mingled with tree-trunks and withered foliage. Amongst these masses, Bates saw many fruits of that peculiarly Amazonian tree, the Ubussu palm, the last he ever would see of the Amazons.

The incredible journey was over and Bates did not undertake anything comparable again. On the surface, his achievements appeared relatively straightforward, but this was far from the truth. During this time he had been predominantly in the company of simple, more often than not uneducated, and sometimes savage people. His regular human contact as he meandered along the rivers was mostly with those he employed as servants and the local people, whom he usually held in high regard.

His journey had been broken down into stages between bases he established along the rivers and it was not until the second visit to Tefé in 1855 that he finally decided to set up a more permanent base. The thoroughness of his exploration was unbelievable and the extent of his preparation is easily overlooked today.

Collecting was always the main purpose of Bates's journeys and he kept meticulous records in pocket books, recording not so much where he was going but what he was seeing. Two of these books are in the collection of the Natural History Museum and are important to view. Written in pencil or Indian ink in the finest script, sometimes minutely small, there is a light pencil sketch of some aspect of the anatomy of an insect. Further on, a drawing has been shaded in by pencil to a remarkable degree of accuracy, and there is an exquisite watercolour of a butterfly, its colours as fresh as if Bates had painted it today. What is clear from these entries is the immense amount of taxonomic knowledge Bates possessed. The logbooks include daily details of the weather, rainfall and mean temperatures, details we would find tedious to keep up today despite modern, easily available sources and a laptop.

Later, as assistant secretary to the Royal Geographic Society, a post he filled for the remainder of his life, Bates advised other travellers about the needs and difficulties they might encounter in their journeys. Few could have been better qualified for the role. He revised and edited instructions for inclusion in a booklet entitled '*Hints to travellers*,' published by the Royal Geographical Society which contains comprehensive instructions to meet most travellers' needs.

Many political, social, and economic changes had occurred since Bates left home in 1848. In 1851 the Great Exhibition had taken place in London, a glorious reflection of the general prosperity and the spirit of optimism then prevailing. This was an era of amazing ebullience and startling contrasts. Strictly speaking, anyone who lived between 1837 and 1901 was a Victorian. In its own time, this was an era of security and national wealth, in which Britannia ruled the waves and foreigners bought British. The principal formative influence, Evangelicalism, had been transformed into a pattern of social attitudes and was greatly strengthened by the example of Queen Victoria's court, where she and her serious-minded consort, Prince Albert, established an atmosphere of moral rectitude remote from the days of the Regency. Church going, family prayers and a general religious propriety were the standard. Coupled with this sense of religious conformity was the Victorians' belief that a man must rely on thrift, hard work and respectability to make his own way through life. *Self-Help* was the title that Samuel Smiles gave to the book that he published in 1859, and the sale of a quarter of a million copies by the end of the century suggests that he spoke for the people of his time.

The Return

Returning to Victorian Britain was a painful dislocation. Bates had after all tried to escape from the humdrum routine of what now had to be regarded as home. He went to Brazil as an inexperienced young man; now he returned mature and nearing middle age. No matter how much he planned for this homecoming, nothing could overcome the fact that he returned to a place barely recognizable. In addition, it would have been understandable if he longed for the hot, dazzling tropics and the freedom of his youth. Initially, Bates attempted to settle down and resume his role as the dutiful son, adjusting as well as he could to his family's expectations for as long as necessary. However, he was unwilling for this to interfere with the sorting and classifying his collections needed. Some of these had already arrived in Covent Garden and he now wanted to prepare the scientific papers on which his reputation would later be judged. Bates arranged to see Stevens in London so that the distribution of his collections could be determined as they arrived from Brazil. He also had to decide what to do about his own reference collection since Stevens was temporarily storing it. Stevens engaged extra staff to help sort, record and catalogue the specimens as they were unpacked.

In the meantime, Bates became involved in the hosiery trade, but it must have then crossed the minds of his family that he might not be as biddable as he first appeared. They may then have even thought he would eventually move to London to be near its scientific societies and his scientific friends. After visiting Stevens, Bates returned home to Leicester and involved himself once more with the family business despite its limitations. The family adapted to these uncertain circumstances hoping that his mother, Sarah, in spite of her years of anxiety and indifferent health would now have a new lease of life. This was not to be for soon after Bates's return she died of a liver disease on 19th January 1860 aged 57.

After Bates's mother's death and as they had no living in servants, Bates's aunt Mary took over the running of the all male household. She was a widow and sister of his father and managed house for them for the rest of her life. It was she who called Bates '*the beetle man*,' a name by which he has been affectionately known by all future generations of his family. The name referred more to his early days in Leicester than to his later time on the Amazons for which he could be more appropriately called '*the butterfly hunter*'.

Sometime during 1860 Bates met 19 year old Sarah Ann Mason. She was a butcher's daughter, from Leicester, and her family was considered socially inferior to the Bateses. Her father was originally a general labourer but the uncertainty of obtaining work made him take a stall as a butcher in the town market in the 1850s. There is no evidence that Henry Bates senior knew Mason nor was Mason associated with the nonconformists. By the time Bates met Sarah, her father was well established and the family made a reasonable living at the trade. There is no account of how Bates met Sarah but it may have been in the market or perhaps she did some domestic work in the Bates's household. Nevertheless, the relationship developed, and steadily became more serious. Sometime during the summer of 1861 Sarah became pregnant. On 2nd February 1862 Sarah and Bates's first daughter Alice was born in Leicester, and out of wedlock.

By the time they married in January 1863 Bates, Sarah and the child had escaped the prying eyes and the gossip of Leicester and were living together in London.

Bates's friend, neighbour, and executor, Edward Clodd glosses over this entire episode in a typical Victorian way in his memoir of Bates in which he said:

> "In January 1861 Bates married a young lady for whom he had kept a tender place in his heart during his long absence [in Brazil]."[126]

This could not have been the case for in 1848, when Bates left for Brazil, he was 23 and Sarah only 7 years old, she having been born on 30th July 1840 in a small house in Fleet Street, Leicester. It is most unlikely Bates would have known or even remembered her, let alone kept a tender place in his heart for a 7 year old girl during his long absence.

When Sarah first met Bates, it is fair to assume she found him more attractive and interesting than the younger less experienced men in her social circle. From her photograph, taken when she was still a youngish woman, she has a pleasant and homely face. The returned explorer must have been a heady cocktail for such a young girl, but his family were less than pleased at the intensity of the relationship. The place on the child's birth certificate for the father's details was left blank and the mother, who was obviously unable to write her name, signed with an *x*.

Edward Clodd said that Bates was married to Sarah by the time the child was born but he is guilty again of manipulating the truth as the birth certificate for Alice is dated 2nd February 1862 and the marriage certificate for Bates and Sarah is dated 15th January 1863. The wedding ceremony took place at the Register Office in the district of Pancras in the county of Middlesex, where bride and groom gave their joint address as 10 Hollis Place, Pancras.

Perhaps Clodd had not been told the true story. But if he had he was no doubt driven by Victorian prudery, as well as a sensitive recognition that an

---

[126] Clodd, Edward, *Memoir. The Naturalist on the River Amazons*. (London: John Murray, 1892).

evasion of the truth gave an aura of respectability to the start of Sarah and Bates's life together.

In marrying her, Bates was behaving as many lower middle class men did when they married girls conspicuously less educated than them. Sarah had already had his child and was no doubt content to fulfil the traditional wifely role. Bates describes himself on the marriage certificate as a worsted hosier, so although supposedly in London for scientific purposes, he still considered himself a tradesman rather than a 'scientist', an occupation that was still evolving from a gentlemanly amateur pursuit to a profession. Bates and Sarah's move to London may have also been influenced by prevailing social attitudes. Perhaps the attitudes of Victorian Leicester proved too much for the couple. Bates's family showed a growing dislike of the relationship, undoubtedly fuelled by the unsuitability of the uneducated Sarah and the continuing scandal of a pregnancy out of wedlock. With a child and unmarried mother, there must have been immense pressure on Bates to do the right thing: marry, and make a new start in life.[127]

With considerable help from Stevens, who was influential in scientific circles and included Darwin and Murray as his friends, Bates's name began to become better known in London. It was nevertheless a difficult social journey, harder in some ways than the physical journey in Brazil, for Bates was excluded from the highest echelons of the scientific elite because of his background.

[127] Lee, Monica, *300 Year Journey: Leicester Naturalist Henry Walter Bates, F.R.S., and his family, 1665-1985.* (Leicester: Castle Printers 1985) 50-51.

Plate 13.
Sarah Bates c 1865-70

This makes Sarah a rather irrational choice for a man who had the clear objective of breaking into the London scientific establishment, a task in which an accomplished wife would have been a considerable asset. Instead, because of an out of wedlock pregnancy, he had acquired a semi-literate wife and it was inevitable she could have found it difficult to socialise in the circles that were so important for her husband's advancement. However, the marriage endured. They had five children, and Sarah grieved greatly for her husband when he died in 1892.

In Victorian times, illegitimacy was not generally discussed, but it happened frequently. There can be no doubt that Bates was Alice's father. She looked like him and inherited his high forehead. He absolutely adored her as men through the ages have doted on their love children. When, as a young married woman Alice died in May 1891, just after the birth of her own daughter, Bates was utterly devastated and was still grieving for her when he died nine months later.

During his time in Brazil, and when he returned home, Bates considered writing a book about his adventures, but quickly shelved the idea feeling there were more important things to be done and there were others better prepared for the task than he was. By nature he did not seek publicity, nor was he particularly reflective. In fact, he was extrovert in his interests and all embracing in his tastes. His reserve was purely over acknowledging his own achievement.

For the remainder of his life, Bates lived in London and, as the family increased, he moved from house to house. Until late 1863, they lived at 10 Hollis Place, where Alice was born. They then moved to 22, Harmood Street, Camden Town where Bates's second daughter Sarah was born on 10th December 1863. This time the birth certificate showed his profession as Author. The Bates's next residence was 40, Bartholomew Road, Kentish Town. The first son, Charles Henry, was born here on 16th August 1865. This time the father's occupation was given as Secretary to the Royal Geographic Society. Bates's second son, Darwin, was born in the same house on 20th January 1867 as was his third son Herbert Spencer on 25th October 1871. Sometime in 1885, probably because of the changing needs of a growing family, the family moved once more, this time to 11 Carleton Road, Tufnell Park, Middlesex. In 1885, they also bought a holiday home or possibly a potential retirement home at Eatsweare Villa, Lennard Road, Folkestone, Kent. Here, Sarah Bates was eventually to spend her widowhood although she died in London.

In 1862, Bates sought employment in science applying for a post as an entomologist in the zoology department of the British Museum made vacant by the retirement of a curator named Adam White. Darwin, William Hooker, and others supported his application but he was found unsuitable through age and lack of formal education and the post went through patronage to the poet Arthur Henry O'Shaughnessy. He turned out to be a disaster as he was more of an etymologist than an entomologist.[128]

---

[128] In the early nineteenth century, the British Museum Library achieved worldwide prominence under Antonio Panizzi. He joined the library staff in 1831, under Sir Henry Ellis, and succeeded Ellis as Principal Librarian (i.e. director of the entire Museum) in 1856.

Panizzi's disdain for the Natural History Department, indeed for science generally, was absolute. This can be seen in his evidence to a Select Committee in 1836. The Committee had been asked to consider whether there ought to be

Maybe it was providential that Bates was refused the appointment as he now set about thoroughly sorting his collections and writing them up for scientific journals as well as preparing the narrative of his travels on the Amazons. As he sold his specimens, they gradually passed into private hands in different parts of Europe. However, a large proportion ended up in the British Museum and the detailed classification and sorting of this component was now undertaken by Bates. Then in March 1860, just four months after the publication of Darwin's *On the Origin of Species*, Bates began to write a series of papers entitled *Contributions to an Insect Fauna of the Amazon Valley*, which were printed in the *Transactions of the Entomological Society* and those of *the Linnaean Society*. At first these were devoted to a new classification of the diurnal Lepidoptera, based chiefly on the structure of the front legs, a classification that was later largely adopted by the majority of entomologists.

Variation fascinated him and he went on to show that, geographically, where no great variety in elevation or natural barriers existed, the effect of minor climatic conditions on the soil influenced the fauna and flora. Bates concluded that the distribution of the butterfly species afforded strong grounds for considering the Guiana region as having specific characteristics, and:

---

more scientists on the Board of Trustees, (the governing body of the Museum). The fifty-one Trustees were for the most part politicians, aristocrats and churchmen with only six scientists among them. Panizzi told the Committee: 'Scientific men are jealous of their authority; they are dogmatical and narrow-minded (they) would spoil the men of rank, and drive them away from the board'.

With this attitude prevailing among the top administrators, it is not surprising that the Natural History Department languished as long as it did. Contempt for entomology in the eyes of the museums trustees is perhaps best illustrated by the annual budget in 1856 for entomological purchases, which was £10.

Another fundamental problem was lack of staff. Until 1856 there were only three Keepers (of Zoology, Botany and Geology-Mineralogy) with a handful of assistants. Places went by patronage rather than by merit, and many a well-qualified scientist was rejected because he did not meet the Trustees' notion of a well-educated gentleman. In 1862 there was indignation when Arthur O'Shaughnessy, nephew of the mistress of one of the Trustees, was appointed Entomological Assistant over the head of the finest entomologist in England, Bates. O'Shaughnessy did not know the difference between a butterfly and a moth; nevertheless, despite protests from the Entomological Society which supported Bates application for the post, O'Shaughnessy's was confirmed (and he was an absolute disaster).

The prevailing snobbery at this time was blatant. J.E. Gray noticed that when he was appointed Keeper in 1840 he was suddenly referred to as John Edward Gray, Esq. (i.e. a gentleman) whereas before he had been plain Mr. Gray. Even as late as 1865, G.F. Hampson, who had been struggling in vain to get a formal appointment in the entomological department since 1860, was suddenly appointed as an assistant when he succeeded to a baronetcy.(Sir George F Hampson) (Barber 163 164).

Bates still hoped for a British Museum appointment and when a new catalogue of butterflies was proposed for the Museum, he thought he would be invited to prepare it. However, the scientific establishment remained impenetrable. It is significant that only the Zoological and Entomological Societies, with their largely amateur memberships, honoured him with Fellowships during these early years. Even the Linnean Society, where he had presented his revolutionary paper on mimetic resemblances, did not recognize him in this manner until 1872, and the Royal Society only accepted him as a Fellow in 1881.

"A very large proportion of species peculiar to itself; the numerous local, sub-species peculiar to the Amazons all showing themselves to be local modifications of Guiana species. (He concluded that Guiana was probably the great centre whence radiated the species which now people the low land on its borders)."

Discussing geographical distribution, Bates gave reasons for dissenting from the then current theories of an extension of the glacial epoch to equatorial regions, whereby many species of the temperate zones would be enabled to pass from the northern to the southern hemisphere. In this, he was considering far more than just differentiation between species, but was reflecting on their origins:

"It is supposed that at that time the climate of the equatorial plains resembled what now exists at six or seven thousand feet of elevation near the equator. It is a tolerably well-established fact that arctic forms then moved twenty-five degrees southward from their homes, and if the decreased temperature then extended to the centre of the tropics, the regions near the equator must have possessed a temperature similar to what is now enjoyed in countries near the twenty-fifth parallel of latitude. Extinction in this case must have been at work largely amongst the forms (if there were any) peculiar to the equatorial zone, and the present character of its fauna ought to show, in consequence, a poverty in endemic forms, and unmistakable signs, in the shape of local varieties or representative species, of a dependence, on the part of the now existing forms, on those living towards the twenty-fifth parallel of latitude; because, with the returning warmth, the extra-tropical species then living near the equator would retreat north and south to their former homes, leaving some of their congeners, slowly modified subsequently by the altered local conditions, to re-people the zone they had forsaken. The present distribution of the species of Papilio does not support the hypothesis of such a degree of refrigeration in the equatorial zone of America, or, at least, does not countenance the supposition of any considerable amount of extinction. The fauna of the Guiana-Amazonian region, so far as regards this genus, is in the highest degree peculiar; showing no dependence on that of the countries near either of the tropics. If now we except the local varieties (the inclusion of which would only strengthen the position), there are about forty perfectly distinct species of this genus inhabiting this region, and of these no less than eighteen are endemic, all of them so peculiarly restricted in their range that they are not found, nor any forms closely representing them, even at twelve degrees of latitude on either side of the equator, The result is plain, that there has always (at least through immense geological epochs) been an equatorial fauna rich in endemic species, and that extinction cannot have prevailed to any extent within a period of time so comparatively modern as the glacial epoch in geology."[129]

Bates was always interested in philosophical matters concerning the greater questions to which his specimens related. In the 19th century, books were the main means of communicating any idea to a wide readership. Controversy was best argued in the pages of a book as in this medium the author was equipped with at least one hundred thousand words to make a point, offering much more scope for success than through letters to newspapers and journals, replies and correspondence. Reviews played an important part too; often written by experts they were the means by which the establishment could legitimately address controversial issues and endorse an author's credibility or otherwise. Books were for that reason very important to Darwin as he saught acceptance for his ideas. What he most needed by 1860 was supporters who through their exploration, observation and writing could sustain his (Darwin's) theories with overwhelming and accurate evidence. Recognising Bates's intellectual ability, Darwin encouraged him to write. Bates sent the first draft of chapter 1 of *The Naturalist on the River Amazons* to Darwin who encouraged him to write more. Darwin, who generously spared time to read the manuscript, wrote to Bates on 13th January 1861:

"I have been very bad for a fortnight, and could not read your MS. before to-day and yesterday. It is, in my opinion, excellent-style perfect, description first-rate (I quite enjoyed rambling in forests), and good dashes of original reflections. I feel assured that your book will be a permanently good one, and that your friends will always feel

[129] *Transactions of the Royal Entomological Society*, (1858) Vol v, 352-353. (Entomological Society as it then was, not gaining is "Royal' prefix until 1933.)

a satisfaction at its publication. I will write when you like to Murray."[130]

Then in 1864, Bates heard there was to be a vacancy for an assistant secretary to the Royal Geographic Society in London and on his second attempt, he secured the appointment. According to the Society record:

"The Committee set to work again [post Greenfield who held the post for less than one year] and eventually selected, though with some misgivings, Bates, who was then in his fortieth year, had never been robust, and was suffering from the effects of eleven years' arduous exploring and natural history collecting in the tropical forests of the Amazons. He was distinguished as an entomologist and had won the warm friendship of Mr. A. R. Wallace, with whom he travelled in Brazil. He was strongly recommended by Charles Darwin, who was struck by his power of mind and ranked him in some respects beside Humboldt. Indeed, he was almost too famous as the author of 'A Naturalist on the Amazons' to be likely to take kindly to a post which had just been reduced in dignity and authority, and which required above everything an orderly mind, patience in routine, and business aptitude. It was only the assurance of his publisher, Mr. John Murray, that Bates, naturalist as he was, was a good man of business too, which induced the Council to appoint him as Assistant Secretary for six months on probation. There never was a happier appointment. It lasted for twenty-seven years, during which the shy and reticent entomologist developed latent powers of insight, judgment, tact, and organization which impressed permanently upon the work of the Society the stamp of solid and unobtrusive efficiency."

Bates's diligence as a natural history collector had been marked. When he and Stevens had completed the specimen sales, Bates was left with a profit of £800, which he wisely invested in booming industrial Britain. This investment generated an income of £40 per annum, a return of approximately 5% net as the financial reward for eleven years hard work and tough living on the Amazons. When considering this sum, it has to be

---

[130] Darwin Correspondence Project, letter 3345, Darwin to Bates, 15th December 1861.

borne in mind that the rent of what could be described as a desirable residence in the suburbs of London was then £40 a year, and in Piccadilly the rent for an unfurnished first floor flat was £60 a year. A private school would educate young gentlemen for between £20 and £30 a year.

Bates had collected 14,712 species, mainly insects, and of these no less than 8,000 were new to science, a record of discovery never rivalled by any other field naturalist in South America. He records:

> "The collections that I made during the whole eleven years were sent, at intervals of a few months, to London for distribution, except a set of species reserved for my own study, which remained with me and always accompanied me in my longer excursions. The following is an approximate enumeration of the total number of species of the various classes which I obtained: Mammals 52, Birds 360, Reptiles 140, Fishes 120, Insects 14,000, Mollusks 35, Zoophytes 5, and Total 14,712."[131]

Bates's claims were to arouse considerable scepticism, particularly when, early in 1863, he made them public for the first time in the preface to *The Naturalist on the River Amazons.* The scientific elite simply did not believe him. Bates had been on the Amazons for more than 11 years, about 4,075 days of which only a proportion, probably 60%, would have actually been spent collecting. Based on that figure Bates would have collected two species new to science every day for the duration of his travels, and in total specimens, almost four each day. It was an extraordinary and praiseworthy effort.

In a letter to William Hooker dated 12th May 1863, Bates says that he is irritated at the attitude of the establishment to his claim for species collected that were new to science:

> "On Monday morning I fell into a nest of hornets at (the) British Museum, in the shape of a knot of the leading curators (Dr. Gray at the head) criticising fiercely my statement of having found 8000 new species out of 14,700."

What Bates was most concerned about was his reputation I should be very vexed if it came to get abroad amongst naturalists that I had

[131] Beddall, Barbara G, ed, *Bates and Wallace in the Tropics: An Introduction to the Theory of Natural Selection,* (London: Collier-Macmillan Ltd, 1969) 139.

exaggerated, but I have not. How could they suppose I should make the statement without a preliminary calculation. The truth is that in all the groups of insects which have been so far well worked out, more than two-thirds of my species were new to science.[132]

It was a sore point for Bates, as his integrity was being challenged, but to prove his point he listed 477 species in this letter of which 324 were new to science concluding with the words:

Out of 14,700 I repeat that in all the obscure small groups of diptera, hymenoptera, moths, etc., of which I sent home some thousands of species, still unnamed for most part, the proportion of new species will be still greater. I mean absolutely new species to Europe; for if you examine Clark's British Museum Catalogue ...you will find that all my new' species did not exist in collection(s) before they were sent home by me.

Having refused to allow any compromise, he gained the support of a number of leading field naturalists, and the dispute finally died down after Darwin and William Hooker expressed their confidence in Bates's figures.

[132] Letter to Joseph Hooker, 5th March 1862, Royal Botanical Gardens Archives, Kew.

Part three

The Evolution of the Man

Chapter 8

Facilitating Science

One hundred and fifty years ago, in one of the greatest revolutions in the history of human thought, Charles Darwin demonstrated beyond reasonable doubt that man was a part of nature and shared origins with all other forms of life. Therefore, existing species, including man, have evolved from common ancestors, and continue to evolve.

The Idea that Species Evolve was not a New One

Wallace had reached a similar conclusion to Darwin. In fact Darwin's own grandfather, Erasmus Darwin, had written a detailed explanation of the idea some fifteen years before Charles was born in the form of a long but unremarkable poem entitled *Zoonomia.*[133] In 1809, the French biologist Lamarck, who remains one of Darwin's most important rivals, published his evolutionary theory.[134] Indeed, even Ancient Greek philosophers had come close to postulating a theory of evolution, but never tested it by experiment or detailed observation. There are reasons to believe that, more than a century before Darwin, Descartes and Buffon were also thinking along these lines – but did not publish their ideas because they feared the likely response of the civil and ecclesiastical authorities. Moreover, a large amount of intellectual debris had to be swept away before there was any possibility of evolutionary ideas gaining widespread acceptance. Darwin was to remove much of this debris and went on to find evidence that could only be interpreted in the way he intended.[135]

Linnaeus was one the most important figures in the development of the idea of species, which was clearly an essential basis for the whole concept of their evolution. Yet, for Linnaeus, species were unchanging God-created entities for man to describe and put in order. Darwin's challenge to the notion of the absolute permanence of species clearly

---

[133] *Zoonomia, or the Laws of Organic Life*, by Erasmus Darwin was written between 1794 and 1796.

[134] Chevalier De Jean-Baptiste Antoine De Monet Lamark (1744-1829) was a pioneer evolutionist who, after the revolution in 1794, was the keeper of insects, worms, and shells at the Musee d'Histoire Naturelle in Paris.

[135] Rene Descartes was an eighteenth century French philosopher who advocated scientific materialism, but in order to escape heresy, decided to postulate a difference in kind between man and beasts, later known as Cartesian duality. Darwin's ideas eventually demolished Descates's thoughts as Darwin showed that all life was related and connected.

implied that traditional ideas of Creation were no longer sustainable. But if species were not unchanging, Darwin faced a new problem, one that would not have troubled Linnaeus. It is obviously harder to define the nature of a mutable species than an immutable one. In other words, the challenge was to identify the attributes that make species fundamental biological entities, even though there may be change over time within them. Here Darwin himself had little to say that was new. In large measure he was prepared to accept the categorization proposed by other competent naturalists; that is those who had wide experience and who, in Darwin's opinion, exhibited sound judgment. This may have been sensible, but it still raises the question of whether their approach to the identification and delineation of species could really provide Darwin with the help he needed to advance his evolutionary ideas from the level of hypothesis to that of theory.

In reality, a satisfactory understanding of the nature of species was unobtainable in Darwin's day. Eventually, it was to be the discovery and development of genetics that would largely solve this problem. Mendel laid the foundations of the science in 1866. Darwin was unaware of this crucial development because Mendel, a monk, worked in virtual isolation and was not allowed to publish. More than half a century was to elapse before genetics reached the point when it could be used to analyse differences between species. Even now the species problem is not completely solved and is still the subject of lively debate.

It is true, however, that even those without formal training in biology can have an intuitive perception of the nature of distinct species. Thus, in most languages, the common names of conspicuous flora and fauna usually refer to what biologists call species. I had a striking demonstration of intuitive perception while in the rain forests of Amazonia. In this region, each square mile contains hundreds of species of trees. Even trained botanists are often at a loss to identify some of the species without examination of their fruits and flowers, and sometimes these are not available. Yet, my indigenous guides and the Amerindians were invariably able to supply me with the vernacular names and uses of the trees after an almost casual examination of the appearance, smell, and taste of the bark or foliage.

In the *On the Origin of Species* Darwin assembled much more factual material of this kind than any previous author had done and could thus put the case for the mutability of species much more convincingly. Above all, he suggested a plausible mechanism that would effect change. Darwin found his clue in Malthus, who had maintained that, in any given region, the human species would always tend to increase to the limit of the food available. But Darwin appreciated that what was true of man was equally true of other species. Most plant and animal species are extremely prolific in seeds and eggs and reproduce much faster than man. But since resources are limited, it follows that death rates are extremely high and a vast proportion of young plants and animals die before they can reproduce.

Darwin argued that, if individuals of any one species vary and if the variations are passed on, it is inevitable that some variant types weaker than their fellows will be more likely to die before becoming parents. Others, the efficient ones, will be more likely to survive and pass their qualities on to the next generation. This is what Darwin called natural selection, a simple and unassailable concept. Given that more offspring are conceived than are necessary to preserve the species, and if there are inherited variations between individuals, natural selection must occur.

Eureka

Darwin's theory was not complete; certainly less so than, say, Newtonian physics. From the beginning, there was a major gap in Darwin's theory. Granted that natural selection caused changes of some kind in living organisms, could it be shown that they actually led to the mutation of one species into another? If it could not, then the whole basis of evolutionary theory would remain flimsy and there would still be room for at least a kind of creationism. The issue would depend on the observation of hereditary variation occurring in wild populations as the raw material for natural selection to work on. However, Darwin lacked the appropriate evidence. This is where Bates enters the story with his paper on mimicry, published in 1863. Bates helped to turn what had still been essentially the hypothesis of evolution in the *On the Origins of Species,* 1859 into the theory expounded in Darwin's next two books, *The Variation in Animals and Plants under Domestication,* 1868, and *The Descent of Man and Selection in relation to Sex,* 1871.

With his theory of mimicry, Bates suggested the development of new species in a convenient and easily understood timeframe. Darwin seized on the idea as proof of evolution. He had at last found the missing link, the essential proof that had hitherto eluded him.

Bates explained the phenomenon in his paper read to the Linnean Society in November 1861. Darwin was not present at the meeting but it is clear that Bates was anxious for his support. According to Emma Darwin's diary Bates, together with Huxley, visited Darwin at Down House and stayed for three days between 18th and 21st April 1862.[136]

The question of mimicry must have featured prominently in their conversations as the visit convinced Darwin of the significance of mimicry. He went on to support Bates's idea with vigour; even acknowledging that Bates had made a better job of explaining it than he could have done himself.

Batesian and Other Forms of Mimicry in Butterflies

Bates's discovery of mimicry and his development of the theory surrounding it represents a 'eureka' moment in science. When scientists talk about a theory they mean they have an explanation, sufficiently confirmed by observation and experiment that experts can accept as fact. This will stand until other severely conflicting data or a better explanation emerges. In other words, a theory is significantly more convincing than a hypothesis that may or may not be confirmed when facts are available.

Bates collected insects extensively, the greater part consisting of gaudy colourful butterflies that sold well in England. Among them were a large number of butterflies that appeared to belong to a group called *Ithomiinae.* These butterflies were small with long slender bodies and eye-catching patterns on their wings. When Bates examined his specimens closely he discovered a few, similar in shape, colour, and markings to the majority, but exhibiting different anatomical features. They were not *Ithomiinae* at all but *Pieridae,* a group to which the familiar cabbage white butterfly belongs. The ground colour in this group is usually white and the shape of the body and wings quite different from the *Ithomiines.* But some of the *Pieridae* now had departed from their family blueprint and had come to resemble the more plentiful *Ithomiines* among which they regularly flew.

Of course, the question was how and why these variant *Pieridae* had diverged from the group norm by copying the external characteristics of a species belonging to an entirely different genus, while at the same time keeping their own genus's anatomical features. Bates decided that there must have been some reason why the variant *Pieridae* became more like the *Ithomiines.* This led him to ask what advantage the *Ithomiines* had over the other butterflies occupying the same territory. The answer was by no means obvious; *Ithomiines* are quite small, rather flimsy and have a

---

[136] Emma Darwin's diary entry for 18th April 1862.

comparatively weak flight. Perhaps the clue was they could hardly be mistaken for anything else.

Bates noticed that *Ithomiines* were rarely attacked by birds despite their obvious availability and apparent vulnerability. He suspected this might be due to their unpleasant taste. He may even have tasted them himself to find out exactly what Nabokov was to do. Bates surmised that the *Ithomiines* conspicuous colour and pattern was a warning, advertising disagreeable properties to potential enemies. A bird that had once attempted to eat *Ithomiines* would subsequently associate their colouring, shape, and habits with an unpleasant taste. After one bad experience any sensible bird would surely leave *Ithomiines* and their copycat *Pieridae*, severely alone. Of course there was no reason for the *Ithomiines* to develop variants themselves. Exact copies were the best way to advertise the group's unpleasantness. It was different for the more palatable *Pieridae*. They might not be able to change their taste but if they could come to look like the *Ithomiines*, they would probably enjoy the same protection. Since the variant *Pieridae* was much less likely to be eaten than non-variants they would become progressively more numerous. Thus the act of palatable butterflies acquiring the outward appearance of unpalatable ones provided the first step in establishing the case for mimicry.

Bates and Wallace examined mimicry in butterflies during their early days on the Amazon. Wallace was later to demonstrate that the same phenomenon occurred elsewhere. The butterflies of the Malaysian Archipelago also revealed remarkable resemblances between species belonging to different families. But Wallace added another dimension. He showed that in some species only the female mimicked a model. There might be several forms of female of the same species mimicking different models and they were always unlike the male of their own species.

Bates realised that his first hypothesis could not explain examples of resemblance where both groups, the mimics and the models alike, were unpalatable. This was clearly the case with the *Ithomiinae* and *Heliconidae*.

All the members of these groups display eye-catching colouring and have an unpleasant taste. It was easy to explain why non-poisonous butterflies evolved to resemble poisonous ones, but what advantage could a distasteful *Ithomiine* gain by mimicking a distasteful *Heliconidae* or vice versa? Bates did not have an immediate answer to this.

However Fritz Műller (1821-1897), a German zoologist and contemporary traveller on the Amazon, provided an acceptable explanation. Young and inexperienced predators probably learn to avoid certain prey through trial and error, via a process of killing and tasting a variety of poisonous species. Obviously if the foul-tasting species varied widely in appearance, predators would have to kill many of each before they learned which to avoid. However, if all the noxious prey looked alike the predator would quickly learn to avoid one basic pattern. In other words the essential difference between Batesian mimicry and Műllerian mimicry is that Batesian involves a palatable mimic looking like an unpalatable model whereas Műllerian involves one unpalatable species looking like another unpalatable species. Műllerian mimicry is common among poisonous tropical butterflies in both the Amazon and elsewhere.

In a locality where there are many models, all with different patterns and colour combinations, all must be tested separately before the unpleasant tasting species are recognised. For example a thousand young birds start their learning on a population of butterflies in which there are five disagreeable species, each with a distinct warning pattern. One thousand of each, or five thousand butterflies in total, will be killed before all the birds have sorted out the taste differences. However, if instead of showing five distinct warning patterns, these five species all displayed the same pattern and colours, the education of the birds will be accomplished at the cost of only one thousand butterfly lives. Even if one of the five species is more abundant than the others all still profit; although the gain is relatively greater in the least abundant species. Műller's views are now widely accepted by students of mimicry as an extension of the original Batesian model.

Műller's explanation depends on the learning capacity of birds. Each year new generations of young birds hatch, and have to learn what is pleasant to eat and what is not. Experience quickly tells them to associate the bright colours of the *Heliconine* and the *Ithomiine* with a distasteful experience and, like their parents, they learn to avoid butterflies that advertise certain conspicuous colour combinations.

First in West Africa and later in Kenya, a century after Bates developed his theory, I observed mimicry in a highly polymorphic species of swallowtail butterfly, *Papilio dardanus,* which also had several mimetic morphs. *Papilio dardanus* probably provides the best example of Batesian mimicry anywhere in the world. The male *dardanus* flies with his harem of different consorts, nearly all tailless, always wonderfully similar to unpalatable forms of other species found in the same localities. This polymorphism is important as it allows a larger total population of *dardanus* without prejudicing the successful mimetic system, as it is essential that each morph remains rare relative to its Batesian model.

Papilio dardanus is endemic to the entire sub-Sahara area, occurring in a number of races in which the females differ greatly in appearance from the males and from each other. The males (1) are bright yellow and have the well-known tail on the hind wings, a characteristic common to most swallowtail butterflies. Some females look like the males and have tails on their wings (2), but others (4 & 6) are tailless. They appear in many different forms and can only be identified as *dardanus* after careful examination. They look remarkably similar to other butterfly species without tails on their wings, usually those inedible by insectivorous predators.

Nine distinct races of *dardanus* have been identified but my illustration only considers two: the female form *hipocoon* from the west coast of Africa and the southern form *triphonius*. A question arises also how sex recognition operates among mimetic species. Is the male programmed to look for his mate in some particular way? At least one of the partners must be able to distinguish the other, despite colour and pattern differences. This probably requires something more than sight and it is assumed that sensory organs and signals must be involved. It is important to establish which senses are employed in each case. For instance, there are distinct differences in the courtship pattern between the mimic *Papilio dardanus* (4) form *triphonius* and its model *Danaus chrysippus* (3).

The model species (3 & 5) belong to the genera *Danaus* and *Amauris* of the family *Danaidae* whose caterpillars feed on plants with a distasteful milk-sap, especially milkweeds and swallowwort (*Asclepiadaceae*). The distasteful plant substances that make the caterpillars inedible transfer to the butterfly, and both caterpillars and butterflies display conspicuous warning colouration.

A *dardanus* male (1) flies underneath the female (2), forcing her to land on his back. The female is then carried downwards on to a bush, and females unwilling to mate have the opportunity to fly off. A male *chrysippus* on the other hand, flies above the female *chrysippus* and dives down towards her. The male attempts to strike the female with his abdomen and then immediately flies upwards. Odours also play an important part in this behaviour. The males of the *Danaus* species have two protrusible, hair-covered odour brushes, or aerial hair pencils, emitting pheromones at the posterior end.

1 Normal male

2 Non-mimetic female

3 Toxic model female

4 Morph A. A mimetic female

5 Toxic model female

6 Morph B. A mimetic female

Plate 14.

The African Mocker Swallowtail

***Papilio dardanus*. Brown.1776 and Batesian models (3 & 5)**

The male bends his abdomen sharply upwards and presses those brushes into special odour pockets on the upper side of his hind-wings. In the courtship phase, this is performed several times a day.

Chemical substances in the brush and in the pocket probably react together to create an odour, a pheromone, which varies from species to species and sometimes from sub-species to sub-species. The male presents the odour to the female when in flight by dragging the brush over the female's head.Female colouration is also an important feature of recognition when males are looking for females of both mimetic and non-mimetic butterflies. My own colour decoy tests in the field prove this. In some cases, particular female colour patterns must obviously be preferred because of the profusion of similarities in nature, but practically nothing is known of the significance of the male colouration for the female. In these circumstances both sexes may employ olfactory orientation devices to supplement visual recognition in identifying each other.

As we have seen, Darwin was impressed by the theory of mimicry first introduced to him in Bates's paper read at the Linnaean Society in November 1861. Bates, Wallace, and Darwin all subsequently concluded that the continuum between forms, races, and species of diversely patterned tropical butterflies provided the only explanation for mimicry and speciation. Modifications to this view in the light of work on reproductive characteristics and reproductive isolating mechanisms has had important implications for the entire biological species concept. [137]

An Ongoing Theory

Even though Bates's ideas have been modified, it is important to stress that their influence was not confined to near contemporaries like Darwin. They also played a central role in the work of two more recent scientists: Cyril Clarke, from Bates's hometown of Leicester, and Philip Sheppard. Clarke and Sheppard bred many hundreds of *Papilio dardanus* in the course of their experiments. They were later joined by Professor E B Ford of Oxford University whose contribution was in the field of

[137] All the material for this chapter is based on discussions in Brazil with William Overal, Professor of Entomology at the Museu Paraense Emilio Goeldi in Pará, Igor Seligman and Andre Cardosos at the same establishment and Dr Everado Martins, butterfly breeder and researcher in mimicry at Alter do Chão, near Santarém, Pará State, Brazil, and on the author's life long study of the subject.

genetics. This work continued until Sheppard died from leukaemia in 1976. By this time two medical benefits from the study had emerged: the demonstration of a link between blood group 'O' in man and the occurrence of duodenal ulcers, and a highly successful immunisation procedure to prevent Rhesus disease in babies.

Clarke began his work while practicing medicine as a consultant physician in Liverpool. His interest in butterflies started during World War I when his parents, fearing that Leicester could be targeted by German Zeppelins, sent their son to the Leicestershire village of Houghton-on-the-Hill. It was there that Clarke's fifteen year old governess, Margaret Fisher, introduced him to butterfly collecting, a hobby he was to pursue for the rest of his life. However, it was his later breakthrough in hand-pairing swallowtail butterflies that made it possible for Clarke and Sheppard to cross races and species of butterflies, thus enabling them to investigate the genetic differences in species.

The first hybrid was between a yellow and black British swallowtail, and a similar black and yellow North American species. Clarke discovered there was only a single gene difference between them, and this was what determined female colouration. Further research into blood groups in man revealed that A, B, and O blood groups are controlled by a similar single inherited gene. This led to the identification of the link between blood group type 'O' and duodenal ulcers.

The extent of polymorphism found in *dardanus* cannot be random, and some form of special genetic arrangement is clearly involved. In *dardanus*, the genes for each different morph are grouped together in a tightly linked complex on one chromosome. For the outcome to be as predictable as it is now, once the mimetic morphs were established, a strong selective pressure must have worked to favour closer linkage between the genes producing the pattern. Sheppard, Clarke, and Ford concentrated on the steps by which wing patterns change in the evolution of a species. The primary question was whether mimicry played any part in this development. As they began to understand the complexity of the subject, they crossbred races of *dardanus* in which the mimetic pattern was present in one but not in the other. They discovered that in these circumstances mimicry was 'broken down', and they advanced a hypothesis that, originally, a major mutant gene had arisen which gave the butterfly some resemblance to a distasteful model. Subsequently during natural breeding, predation would occur but its effect would be lessened over an extended period by natural selection. Minor genes would therefore gather on the same chromosome leading to a continuous refinement of the mimicry resemblance.

The complete set of mimetic genes is known as a super gene and would be inherited as a unit. Further experimentation showed that 'tails' on the wings of the swallowtail butterflies, could be changed from being inherited with a yellow colour, to being inherited in black - technically described as a crossover. This proved crossovers could be artificially bred in the same way as they appeared occasionally in nature. As a result, the theory was considerably strengthened.

Experiments continued with a number of different species of butterflies, including the North American *Papilio glaucos* of the same family as *dardanus*. The inheritance pattern in the wings of these swallowtails encouraged Clarke, Sheppard, and Ford to reconsider blood groups, this time the Rhesus system. They became convinced that a super gene inherited the genes controlling the colour and wing patterns in butterflies in much the same way as the same mechanism controlled Rh in man. The three scientists discovered a method of preventing Rhesus disease, a treatment that brought Bates's mimicry, as their starting point, back to the centre of science and medicine. The trio developed the *Liverpool jab*, given to a Rh-negative mother after the delivery of a Rh-positive baby. The injection, an anti-D gammoglobulin, is now standard practice in obstetrics. It destroys any Rh-positive cells that have entered the mother's circulation, preventing the cells from immunising the mother and causing problems in subsequent pregnancies. For his work as leader of the team Cyril Clarke was made a Knight Commander of the British Empire in 1974.

The core of evolutionary theory is that all life forms are connected to each other through a common ancestry. Today molecular biology has reinforced this view to a far greater extent than was thought possible during the lifetimes of Darwin, Wallace and Bates. One of the major developments in modern evolutionary biology is a methodology to estimate the underlying patterns of ancestry among living things. Today molecular data is used to estimate the common ancestries of life, for example between bacteria and the energy-producing nucleus of human cells, intermediate ancestries, and in tracking infectious diseases.

The Key to Natural Selection

By the 1860s Bates knew that mimicry was the key to natural selection. His interpretation of mimicry was that an aposematic inedible model had an edible mimic. The model suffered by the mimic's presence because the aposematic signal aimed at the observer was diluted as the chances increased that the observer would taste an edible individual and fail to learn the association between aposematism and distastefulness. The mimic gained from the presence of the protected model and from the deception of the observer. The observer benefited by avoiding the noxious model, but missed a meal through failure to recognise that the mimic was edible. These mimicry relationships hold up as long as the mimic remains relatively rare. However should the model decline or the mimic become abundant, the protection given to the mimic by the model will wane because the naive observer increasingly encounters and tastes edible mimics.

Vladimir Nabokov comments on mimicry in *Speak Memory*:

> "The mysteries of mimicry had a special attraction for me. Its phenomena showed an artistic perfection usually associated with man-wrought things. Such was the imitation of oozing poison by bubble-like macules on a wing (complete with pseudo-refraction) or by glossy yellow knobs on a chrysalis. When a certain moth resembled a certain wasp in shape and colour, it also walked and moved its antennae in a waspish, un-moth like manner. When a butterfly had to look like a leaf, not only were all the details of a leaf beautifully rendered but also markings mimicking grub-bored holes were generously thrown in. Natural selection, in the Darwinian sense, could not explain the miraculous coincidence of imitative aspect and imitative behaviour nor could one appeal to the theory of the struggle for life when a protective device was carried to a point of mimetic subtlety, exuberance, and luxury far in excess of a predator's power of appreciation. I discovered in nature the non-utilitarian delights that I sought in art. Both were a form of magic, both were a game of intricate enchantment and deception."

Without Bates's theory of mimicry, it would have been difficult for Darwin to persuade his readers that the idea of evolution was valid or that it operated by natural selection. By 1863 Darwin had fullty accepted the concept that natural selection was driven by variation and was best illustrated in mimicry. Bates was to help Darwin to recognise this; he was to demonstrate remarkable intellectual capacity both in appreciating the relationship between variation and mimicry and to explain it in readily comprehensible way. What is thrilling today is that these early ideas are still being bolstered by modern science, such as genetics, of which Bates and Darwin had no inkling.

Plate 15.
A moth or nut husk?

Chapter 9

Facilitating Geography

But now, early in 1864, because of the pressing need to support a growing family by adding to his investment income with a regular salary, he took up his appointment at the Royal Geographic Society.

The Geographical Society occupies a unique place in the cultural history of exploration. Its origins lay in a resolution passed at a meeting of the Raleigh Travellers' Club, chaired by John Barrow, permanent secretary at the Admiralty, on 24th May 1830. Their aim was to form a new Society for the promotion and diffusion of that most important and entertaining branch of knowledge, Geography. Geography was important because it was useful, both to science and to government, and it was entertaining because it provided satisfaction to those prepared to study it.[138]

From 1830 to 1840, the society met in rooms at the Horticultural Society in Regent Street, but by the time of Bates's appointment in 1864, it had moved to 15 Whitehall Place. In 1870, the Society found what it then thought

---

[138] Sources consulted were: Corr. block 1861-70 Two letters: 16th July 1864 to Capt George, Royal Geographical Society Map Curator on the progress of publication of Baines 'Illustrations of the Victoria Falls'. 13th July 1865 to Captain George, on the preparation of maps for Capt Toynbee and Samuel Baker.

Corr block 1871-80 Five letters: 3rd October 1872 to Francis Galton asking for sanction for printing of R F Burton's translation of Lacerda's journey. 27th August 1872 to Farley, clerk at Royal Geographical Society, instructions on calling a meeting of the Livingstone committee. 28th August 1872 to Farley also concerning the above. 11th January 1875 to Royal Geographical Society Secretaries resigning Fellowship of Royal Geographical Society. 28th July 1880 to Evis, clerk at Royal Geographical Society.

Corr block 1881-1910 Five letters: 28th August 1882, draft of letter agreeing to loan of instruments for polar station at Fort Rae. 29th May 1884 to Francis Younghusband on sources of information about Tibet. 7th September 1890 to Evis, forwarding letters needing attention while he is on holiday. 3rd March 1891 to Capt Oliver on a proposed bibliography of Madagascar. 6th February 1892 to J S Keltie (Bates's successor) has influenza and must stay at home.

Miscellaneous: Two letters from Mrs. Bates after death of her husband, 1892 and 1897.

Invitation from Fishmongers Company, 1890. Invitation card from Mr. and Mrs. Fred Low, 887. Letter and note on Bates's grave, from Dr J P Dickenson 14th June 1989. 6th October 1948, letter from H P Clodd forwarding a pincushion that was used by Bates during his Amazon travels. Duplicated article by Ricardo Ferreira 'Henry Walter Bates and the controversy on evolution and group selection'.

Scott Keltie correspondence: four letters 1884-85 about Keltie's appointment as Librarian of Royal Geographical Society. Back Collection: two letters, 1875, replies to invitations. H R Mill Collection: (No. 5f) 4 letters, 1887-1891 on various articles submitted by Mill for publication in the Proceedings. Referee reports with Journal MS Islands 1880 (A H Markham); Journal MS Central Asia 1869 (Semenoff), Journal MS Japan 1874 (Jeffreys), Journal MSS Central America 1868 (Cockburn) nos l& 2, Journal MSS South America 1870 (Chandless) 1876 (Bigg-Wither); 1877 (Simpson) no.2 1880 (Christison) 1880 (Crevaux), Keltie J S news cutting commenting on Bates in Keltie corr.

Hooker J D letter of 25th March 1892 comments on Bates.

Nery, Baron Santa Anna letter of 8th March 1892 refers to death of Bates.

Werner, J R Note by Bates on back of Werner's letter of 30 Mar 1891.

Wallace A R letter 8th March 1892 comments on Bates.

Clodd H P letter of 1930 about Bates's Amazon notebook.

Needham J F corr. 1881-1910 includes draft of letter from Bates's Royal Geographical Society Letter books.

would be its permanent home and moved under Bates's supervision to 1 Savile Row, an address that quickly became associated with adventure and travel. The Society also used a lecture theatre in Burlington Gardens lent by the Civil Service Commission, although this accommodation was found to be unsatisfactory. Finally, in 1913, it established a permanent home at 1 Kensington Gore, as a world renowned centre for geographical learning, research, education, and planning expeditions.[139]

Like many learned societies, it began as a dining club where interested members held informal dinner debates and discussed current geographical and scientific issues. In 1859, the society changed its name to the Royal Geographical Society on being granted a Royal Charter by Queen Victoria.

Bates first applied for the post of Assistant Secretary to the Royal Geographical Society in 1863. The unexpected resignation of the Honorary Secretary, Francis Galton, forced the President, Sir Roderick Murchison, to plead with the Council to look for a replacement swiftly. A Mr. Greenfield was appointed as assistant secretary in preference to Bates but Greenfield died within a year of taking up the appointment.

Bates successfully applied again in 1864, this time supported by Darwin and Murray. The only other applicant was Wallace.[140]

The Appointment

After lengthy discussions among Council Members Bates was appointed on probation. Some members of the appointment committee still had reservations. Bates was in his fortieth year, too old perhaps for the pressure and stress of the appointment especially in the light of Greenfield's sudden death. There were genuine concerns about Bates's health. The committee feared that Bates's years of demanding travel and exploration had taken their toll. After Galton, the job had been scaled down and there were fears that Bates, the author of *The Naturalist on the River Amazons*, might be too famous for a post that carried less dignity and authority than in Galton's day. However, Darwin's support for Bates was probably decisive. Not only did Darwin attest to Bates's intelligence, he even ranked him in some respects as equal to Humboldt – in short, a very strong recommendation indeed.

The qualities most needed were those of an orderly mind, patience in routine and business aptitude. Here Murray, who described Bates as a good

[139] Driver, Felix. *Geography Militant.* (London: Blackwell, 2001) 46-48.
[140] Moon, H.P, *Henry Walter Bates FRS, 1825-1892: Explorer, Scientist and Darwinian.* (Leicester: Leicester Museums, Art Galleries and Records Service. 1976) 62-64.

man of business, was helpful. Finally, on 25th April 1864 the Royal Geographical Society Council appointed Bates on six months probation. His 'salary to midsummer' was £75, which Bates accepted with alacrity. Though slightly chagrined, the unsuccessful Wallace admitted that Bates was the best man for the job. The decision to appoint Bates proved extremely wise, as he was to remain the Assistant Secretary until his death in 1892. During those years, the previously shy and reticent entomologist developed impressive powers of insight, judgment, tact and organisation. The work of the Society ran with striking efficiency and Roderick Murchison was to comment that the conciseness and grace of Bates's literary style gave new precision and charm to its publications.

As Assistant Secretary, Bates was responsible for the administration of the Royal Geographical Society's affairs, including the organisation of meetings, supervision of the office and all worldwide communications. He also edited the Society's Proceedings and Journal and drafted the annual review on the state of geography for the presidential address. Much of this basic administration was to do with exploration.

Different people, including naval officers, soldiers, naturalists, artists, surveyors, missionaries, novelists and journalists, gathered exploration knowledge in different ways and for different purposes. Many performed several of these roles at once. Livingstone, for example, simultaneously acted as a missionary, roving consul and explorer, whilst his nemesis Stanley was primarily journalist, conquistador, geographer and sometimes masquerader. The task the Geographical Society had set itself in 1830 was to co-ordinate the geographical knowledge produced within the diverse and expanding field of exploration. This provided the grounds for the establishment of an institution that would manage the collection, storage and dissemination of geographical knowledge in a coherent way. The co-coordinating role of the assistant secretary had three essential features:

1. The construction of guidelines for the recording of topographical and other geographical information in the field, according to standardized observational procedures and rules of measurement.
2. The establishment of a centralized archive of authoritative geographical information, available to legitimate explorers and others to whom such knowledge would be useful, including departments of government.

3. The diffusion of geographical knowledge in a rational and educative manner.

In the mid-nineteenth century, through events like its African Evenings, the Society became associated in the public mind with many of the most notable expeditions of the nineteenth century (viz Livingstone and Stanley), and its legislative body was keen to make the most of the publicity this generated. With limited assets allowing relatively meagre direct support to expeditions, especially when compared with the role of the Royal Society or the Admiralty, it nevertheless played a key part in debates over the conduct and significance of exploration. Its publications reported the results of global exploration. Its officers including Bates advised prospective explorers about how to investigate unknown territories and its leading figures like Murchison claimed to exert considerable influence in the corridors of Whitehall. This high public profile during the mid-Victorian years resulted in a flourishing society, making it the largest scientific Society in London by 1870, with a membership of seven thousand.[141]

There were however critics within the scientific and scholarly communities who found fault with its success, regarding the Society's attempt to court publicity with growing scepticism. These anxieties may have been well founded for, from the beginning, it was a hybrid institution which simultaneously sought to acquire the status of a scientific Society and to provide a public forum for the celebration of a new age of exploration. In addition, the hybrid character of the Society – part social club, part learned Society, part imperial information exchange and part platform for the promotion of sensational feats of exploration – frustrated those seeking more rational approaches to modern scientific endeavour.

One severe critic within the scientific community who complained that the name of science was being ruined by the Geographical Society's craving for sensation was Joseph Hooker, the director of Kew. In expressing his concern in 1864, the year Bates's tenure started, he wrote:

I hate the claptrap, the flattery and the flummery of the Royal Geographic, with its utter want of science and craving for popularity and excitement, and making London Lions of the season of bold Elephant hunters and Lion slayers, whilst the steady, slow and scientific surveyors and travellers have no honour at all.[142]

---

[141] Driver, Felix. *Geography Militant.* (London: Blackwell, 2001) 69-89.

[142] Stafford, *Scientist of Empire*, 59.

However his views did not stop Joseph Hooker from accepting a founder's medal from the Royal Geographical Society in 1883.

On the other hand, a significant proportion of the membership were scholars and antiquarians, many of whom found themselves derided as mere armchair geographers by the sensation seekers and African sympathisers. Under Bates's stewardship the society, which was becoming a commercial enterprise as much as a learned society, capitalised on the sensational publicity surrounding African exploration. In the process it extended its operations well beyond the boundaries of either science or scholarship. This was the key to expanding its membership which occasionally compromised its learned society status leaving it as we see it today, a hybrid organization attempting to represent diverse interests behind the façade of a scholarly establishment.

Meanwhile distinct inner circles at the heart of the society emerged represented by dining clubs like the Kosmos. These were a response to the almost unmanageable size of the Society and the inevitable effect this had on scholarship. They enabled small coteries within the society to preserve distinct spaces for maintaining their unambiguous academic interests. Bates, for instance, was eventually to become a leading figure in the Kosmos, which had been founded in 1850.

Bates also acted as secretary to the geography section of the British Association for the Advancement of Science, (BAAS), and was involved in planning its programme between 1864 and 1891, serving as vice-president in 1887 and 1888. He was a member of a number of the Association's commissions including those concerning Morocco and New Guinea, and of all those dealing with geographical education.

The Geographers and Explorers

Between 1864 and 1892, the Royal Geographical Society distributed more than £20,000 specifically for African exploration. This was a substantial sum even allowing for the fact that Britain was developing a strong presence in the Afro-tropical region. The Society was entering one of its most interesting and active periods and Bates became involved with many famous names. During the nineteenth century the Society was visited by many reputable or would be explorers and, after 1864, all would have sought out Bates. As Clodd remarks:

> "Every traveller, both English and foreign, who landed on our shores made his way to [the Society to see] Bates."[143]

Indeed little happened in either physical or academic geography without Bates's involvement. Bates's own journeys provided an admirable model for these travellers and he gave them excellent advice on how to investigate geology, climate and biology. Some of this advice was included in the 1871 edition of the Society's *Handbook to Travellers* edited by Bates. Most explorers clearly regarded the book as important and carried it with them on their journeys.

When travellers returned and reported to the Society, Bates proved a skilful editor of accounts that were to appear in Royal Geographical Society journals. He also found time to edit several travel books, including Belt's *The Naturalist in Nicaragua* (1874), Humbert's *Japan and the Japanese* (1874), Koldeway's *The German Arctic Expedition of 1869-70* (1874), and Warburton's *Journey across the Western Interior of Australia,* as well as a six volume compendium of *Illustrated Travels* (1869). Many of these works contain generous tributes to Bates's assistance. Belt thanks Bates for 'undertaking the superintendence of these sheets in their passages to the presses'. Warburton explains that his travels in Australia had left him so exhausted that he was 'physically unfit to prepare my journals for publication'. Hence, they had been edited by Bates – although with an historical account by Charles Eden and inputs from Professor Owen, Dr J. Hooker, John Gould and Dr Henry Trimen. All of these publications were personal revenue earners for Bates.

During Bates's time with the Society geography itself was changing. An earlier emphasis on exploration was giving way to a new concept of geography as an academic discipline with a rigorous scientific base and also as a suitable subject for study in schools and universities. Bates did much to promote this view and played a significant part in this debate, working hard behind the scenes to make Geography more scientific, academic, and better understood by a wider audience.

A group of reformers within the Society decided that a Chair in Geography should be established at either Oxford or Cambridge in order to promote the study of geography in schools, to train Geography teachers and to give the subject greater academic respectability. On 17th March 1884, the Scientific Purposes Committee of the Royal Geographical Society urged

[143] E Clodd, *Memories* (London, 1916) 68.

Council to appoint an Inspector of Geographical Instruction for one year. The Inspector was to investigate the state of Geographical Education and to make a collection of the best teaching aids then available. In June, James Scott Keltie was appointed to this task and subsequently received a set of written instructions, which included asking him to look at Geography teaching on the Continent and to find out whether it was taught there as a discipline or as in virtually all English schools simply by rote.

The Society ran a series of highly successful prize examinations between 1869 and 1884 and a programme of scientific lectures on physical geography between 1877 and 1879. In the debate on the nature of Geography Bates was very much the modernizer and, along with Murchison and Markham, played an important part both in raising the profile of the Royal Geographical Society and in securing Geography's position in English education. Though perhaps less spectacular than his achievements in Brazil, his work with the Royal Geographical Society may have had even greater long-term significance. He was quick to recognize the potential of the young Halford Mackinder, a superb lecturer, and later one of the most effective champions of academic geography. At Bates's invitation, in January 1887, Mackinder gave a lecture to the Society entitled *The Scope and Methods of Geography.* Later the same year, Mackinder's appointment as the first Reader in Geography at Oxford marked the 'arrival' of the subject in the British academic world. The Readership at Oxford was partly funded by the Society and, as Mackinder acknowledged, Bates played a major role in establishing the post.

Wallace and others later claimed that Bates's duties at the Society were so onerous that they precluded him from pursuing his work as a naturalist. Yet, while the Royal Geographical Society certainly occupied most of his time, Bates continued to produce an impressive flow of entomological papers; the earlier papers were based on his own collections while later ones investigated insects sent in by travellers from around the world. These papers were published in the *Transactions of the Entomological Society*, the *Entomologist's Monthly Magazine*, the *Annals and Magazine of Natural History,* and other journals. Bates also made a major entomological contribution to the *Biologia Centrali Americana* by F. D. Godman and O. Salvin (1880-1911). However, after his mimicry paper, he wrote little on Darwinism or on the geographic distribution of species. Bates's scientific standing was acknowledged by his election to honorary fellowships of the Zoological Society in 1863, the Entomological Society in 1871 and the Linnean Society in 1872. He served as president of the Entomological Society from 1868 to 1869 and again in 1878 but then resigned after one year because he believed that his other commitments (to geography) prevented him from devoting as much attention to the Entomological Society as he would have wished. The entry in the Royal Entomological Society record shows that:

> "Bates influence on the Society was, in the nature of things, rather indirect, but nevertheless his interest was great, and, although by profession his work lay in another sphere, he was predominantly an Entomologist, and a great deal of his very limited spare time was devoted to his favourite study. As a systematist, much of his work was done on the Coleoptera, but the observations he made on the Amazons caused him to look far beyond mere specific differences, as is evidenced by the theory of mimicry, which bears his name. He was elected a fellow of the Royal Society in 1881 [regarded today as the highest accolade a scientist can receive next to a Nobel Prize]."[144]

In the Archives of the Royal Geographical Society there is a poignant letter dated 6th February 1892. It is Bates's final communication with the Society for which he had done so much. Writing only ten days before his death, he is still concerned about the administration of the Society's affairs.

> Feb. 6. 1892,
> Carleton Road,
> Tufnell Park, N
>
> Dear Keltie,
>
> My malady has taken a bad turn and the doctor says it is influenza. He orders me to be confined to my room for a week or two. I am afraid you will have some difficulty with Younghusband's paper which will have to be shortened fully one half and he has not said anything on the subject.
>
> I hope you will take my place at the Kosmos Club.
> Yours sincerely, H. W. BATES.[145]

---

[144] *The History of the Entomological Society of London*, 1833-1933, (London: RES 1934) 143-144.

[145] Moon, H P, *Henry Walter Bates FRS, 1825-1892: Explorer, Scientist and Darwinian.* (Leicester: Leicester Museums, Art Galleries and Records Service. 1976) 64.

He died of influenza at home on the 16th February 1892.

Obituaries

Testimonials given after a sudden and unexpected death are often especially revealing. There were meetings of the Royal Geographical Society and its Council on 22nd February 1892 when the Council was informed of Bates's death:

> "It is with the deepest regret that we have to announce the death, at the age of sixty-seven years, of Mr. H. W. Bates, FRS, of bronchitis following on influenza. He was for twenty-seven years the able, universally esteemed, and much-loved Assistant Secretary of the Society and Editor of its Transactions. For the present it is impossible to do more than make the sad announcement, and refer to what is said about Mr. Bates in the report of the last meeting of the Society."[146]

At this meeting J. Scott Keltie, Librarian for the previous seven years, was appointed to succeed Bates as Assistant Secretary and editor of Transactions. The President, Grant Duff, declared:

> "Before proceeding to the special business of this evening, it is my painful duty formally to announce to you what most of those present have already learned from the newspapers, that our excellent Assistant Secretary, Mr. Bates, died on Tuesday last. Since Mr. Bates first connection with the affairs of the Society, more than a quarter of a century ago, he has had thrown upon him an increasing amount of varied work, and he has done that work always admirably well. Every President, every Honorary Secretary, every Member of Council, and every officer of the Society of every grade, has found in Mr. Bates an admirable co-operator and an admirable friend. He was so good an official that when one talked to him one was apt to forget that before he began to be an official, before he began to be a professional geographer, he was a most accomplished naturalist and the author of one of the best books of its kind in the English language."

[146] Royal Geographical Society Archives, Geographical notes, NS 14 182, 177.

The President went on to say that all who met Bates found him helpful, patient and the most amiable of men. He paid tribute to Bates's skill when editing the thoughts of accomplished and able travellers who had never practiced the art of compressing what they had to say into the brief scope of a paper to be read at an evening meeting. Duff asked the Council to express their condolences to Bates's widow and continued:

> [The Society]..."trust that the knowledge of the estimation in which Mr. Bates was held by all who knew him, and above all by his colleagues and those best able to judge of his scientific attainments and high and generous character, may be in time a consolation to his family. They feel that the loss to the Society is one which can hardly yet be estimated, and they desire to tender to Mr. Bates family the assurances they hoped to have offered to himself, on his retirement, of their most warm and heartfelt appreciation of the devotion to the interests of the Society shown by him during his twenty-seven years work as Assistant Secretary and Editor of Transactions."[147]

Lord Aberdare gave a further tribute:

> "He was one of the rarest characters I had ever known. Considering the vastness and variety of his knowledge, it was astonishing to find a man so gifted, with such entire self-effacement and modesty. You may well believe that the office of President, especially to one like myself, who had only a general and superficial knowledge of geography, is not merely difficult but impossible without the assistance of the standing officials; and in Mr. Bates I found not only an ardent follower of knowledge but one of the most sagacious of men. He knew men as well as he knew the butterflies, to seek which he first made his acquaintance with the Amazons. He was a great reader of human nature, but he was more than that. We all of us in the course of our lives, I hope, have met many men who have commanded our respect and our regard. Bates was something more than that. It was impossible to associate with him without feeling not only regard but personal affection; and so it is that all those who are taking part in this ceremony to-day feel that in him they have lost one of their best and

[147] Royal Geographical Society Archives, report of the Evening Meetings, NS 14 182, 10.

dearest friends. This is but a small tribute to his memory and to his worth, but short as it is I could not but offer it."[148]

Mr. Douglas Freshfield then added his comments on his friend:

"It may seem superfluous to add anything to the fitting tributes already paid by you, Sir, and by our late President, Lord Aberdare, to the memory of the colleague and friend we have lately lost. But I feel assured that our Fellows and our guests present here to-night will recognise that to remain silent would be unnatural, if not impossible, for me, who have in the course of the past ten years, in the conduct of the affairs of the Society, been brought more continuously and more intimately perhaps than any one else in contact with Mr. Bates."

He praised Bates as a scientific explorer of infinite patience, endurance and enterprise, as the writer of one of the best books of travel ever written, as a naturalist, close observer, and as a writer of style envied by Darwin. He continued:

"As the head of our office, Secretary, and Editor of Transactions, a very large responsibility rested on his shoulders. He was brought into contact with all the great travellers and many of the little travellers who visit our capital. He had to tame and manage the last African explorer, or to draw out what was worth drawing from a timid missionary... With each and all, Bates got his own way in the end and that without quarrelling. He was a very shrewd and by no means always lenient observer and judge of men and manners. His judgments were deliberately arrived at, and expressed with a commanding modesty. But he was essentially generous; it was his pleasure to elicit and bring forward the best in all he had to deal with, men or manuscripts...He would re-write...an obscure and tedious paper with as much pains as he would bestow on the material of a Presidential address."

Freshfield noted that the Society's work had changed greatly during the last decade. Bates had turned its transactions into a monthly magazine and

---

148 Royal Geographical Society Archives, report of the Evening Meetings, NS 14 182, 11.

assumed the editorship; he had striven constantly to widen the Society's scope and strengthen its links with Commerce. Freshfield concluded:

> "All these ... have meant, in one way or another, new work for the Secretary. Bates never shunned work. Each new proposal he turned over carefully, and criticised independently and usefully, but once accepted in his mind as within the scope and resources of the Society, he carried it out with unflagging energy."[149]

A note from Clements Markham to Freshfield explains what he would have wished to say himself before the Society:

> "I should like to have said something at the meeting last night to express my feelings about our dear old friend Bates, but there did not seem an opportunity ...I always remember Bates great modesty even more than his judgment, which was excellent and his goodness and kindness of heart, which never failed. I knew him when he first came home from the Amazons, and he dined with me twice in 1863; I was charmed with his conversation and his thoughtful way of dealing with every subject. Norton Shaw was then going away, and I suggested to Bates that he should apply for the Assistant-Secretaryship; but he was diffident about it and said he had no experience, though there was nothing he should like so well. Mr. Greenfield was appointed, but after a short tenure died very suddenly in our house in Whitehall Place. Meantime I had spoken about Bates to Sir Roderick, who had seen him and been extremely pleased with him. The ground had thus been prepared and Bates was appointed. I cannot tell you what help he was in those days when everything was in utter confusion...no one in the Library but a negligent boy, and tremendous work to get things into order, Bates doing the lion's share. Sir Roderick's high opinion of him increased every year. I mention this that all may know Bates was not a man to push himself, but on the contrary we sought him out. I think this modesty was carried to a fault; for if he had had less we might have had much valuable work from him in the way of views and deductions on geographical points. We shall never see his like

[149] Royal Geographical Society Archives, report of the Evening Meetings, NS 14 182, 12.

> again, and those who have known him longest will certainly not feel his loss least."[150]

The Officers and Members of the Society remembered Bates both with affection and respect for his professional role as the Society's administrator. To them he was first and foremost a Geographer rather than naturalist of the Amazons.

Other Archive Material

There are some interesting unpublished papers in the archives of the Society. One is the certificate relating to Bates's application, made on 8th January 1866, seeking election as a Fellow of the Royal Geographical Society. Bates may have felt that a Fellowship would have confirmed his standing within the Society and given him greater authority when dealing with outsiders. The Royal Geographical Society rules stipulated that such applications should be signed by at least two Fellows. Bates's application carries no less than eighteen signatures. Livingstone's application is the only other certificate in the Society's records with so many. In both cases, the extent of the support testifies to the esteem in which the candidate was held. This must have been a testament to the praiseworthiness of the candidate in the eyes of their peers. Membership was obviously important to Bates, yet nine years later, in 1875, he wrote a letter to himself as Assistant Secretary tendering his resignation:

> "Having found it necessary to discontinue my subscription kindly take off my name from the list of Fellows from 1st of January of this year. [1876]"[151]

In view of the widespread support Bates had received at the time of his application and of the clear signs of an ever-rising reputation in the intervening period, his resignation seems surprising. Unfortunately, there is nothing in the Royal Geographical Society records to shed any light on this event. It is just possible that Bates decided that he could no longer afford the annual subscription of £2. This seems unlikely and, even if it were true, one might have expected the Society to waive the subscription of a man who was doing so much good for it. Bates may not have been rich but it is interesting

---

[150] Royal Geographical Society Archives, report of the Evening Meetings, NS 14 182, 13.
[151] Royal Geographical Society Archives, CB6 (1871-1880) Bates, HW.

to note that his estate at probate was worth £7,974 7s 3d, a tidy sum in 1892, the equivalent today of more than £800,000. It is perhaps more likely that Bates resigned on the principle that he was tired of the struggle between membership, popularity and scholarship and was making a point of it.[152]

A friendly association between Sarah Bates and the Society is suggested in two unpublished letters written for or by Sarah Bates after her husband's death. The first, written in a mature hand and well-educated style, was probably by Edward Clodd, who delivered the Society's condolences to Mrs. Bates. It reads:

> "Dear Mr. Keltie, The copy of extract from the minutes of Council of the Royal Geographical Society held on the 14th inst, has been handed to me by W. Clodd, and I beg that you will convey my grateful thanks for the substantial recognition of the long services of my dear husband; which they have been pleased to make to me. Will you at the same time express my heartfelt thanks for their sympathy with my irreparable loss, which was conveyed to me in their resolution of 22nd Feby. I remain, Yours very truly, S A Bates."[153]

The second letter is more revealing and written in an immature hand and style. I suspect that this letter was actually written by Sarah Bates herself. It has been widely supposed that Sarah was completely illiterate, yet it is surely possible that Bates had taught her the rudiments of reading and writing. The letter is certainly not the work of the previous letter writer:

> "My Dear Mr. Keltie, You perhaps have heard from my daughter that I am about leaving London I have tried hard for the last 5 years to make things meet and tri as I could not let my Folkstone house to make any money. Whatever of it so I have made up my mind to go & live there as I am quite alone in the world now & do not go out anywhere or into any Society whatever since my husbands death I sincerely hope that you Mrs. Gilmore & Family are all quite well will you please give my kind regards to Sir h Johnston when you next see him. I have so longed to come to the Society but feel I could not endure it without my husband will you please forward next magazine

---

152 Dictionnary of National Bigraphy. Also Henry Walter Bates & Lee, Monica, *300 Year Journey: Leicester Naturalist Henry Walter Bates, F.R.S., and his family, 1665-1985.* (Leicester: Castle Printers 1985).

153 Royal Geographical Society Archives, CB7 (1881-1910) Bates, SA..

to Folkstone with kind regards to yourself kind remembrances to all enquiring friends Believe me yours truly S A Bates."[154]

Most commentators on Bates present him as a modest and unassuming man. This is certainly the impression given by a letter in my possession. In it, Bates was writing to an unknown correspondent who had asked for his autograph. The date suggests that he was both known and popular quite soon after his return from the Amazons:

22 Harmwood Street,
Haverstock Hill, NW

23 May [1864]

Dear Sir,

I thank you heartily for the kind mention of my book of my travels in your note & as the honour is not small of having one's autograph classed with those of the many distinguished persons in your list [I] gladly comply with your request by sending mine
Yours sincerely,
H W Bates.[155]

Bates was a self-effacing man who shunned celebrity, but various continental and American learned societies appointed him to corresponding membership, and the last Emperor of Brazil, Dom Pedro II, made Bates a Chevalier of the Order of the Rose, a decoration that he did his best to conceal whenever he wore it. At one Society event, Galton caught a glimpse of the insignia and wanted to take a closer look, but characteristically it was kept hidden beneath the lapel of Bates's jacket. Even so, only the most insensitive of men and Bates was far from that could have failed to be gratified by such an honour accorded to few non-Brazilians. It appears that even Bates's family was unaware of this honour, as he never showed it to them. There seemed to be some doubt about the Order of the Rose, but I have traced a record of the award in the papers of Dom Pedro II's private secretary.[156]

---

[154] Royal Geographical Society Archives, CB7 (1881-1910) Bates, SA..
[155] Author's private collection.

In March 1892, the Foreign Secretary of the Royal Geographical Society received a letter from Baron de Santa Anna Nery, an Honorary Corresponding Member of the Society. The Baron wrote from Paris of his grief at the news of Bates's death:

> PARIS: 66, RUE MOZART, Le 8 Mars. (1892)
> My Lord,
> On my arrival from Brazil, where I have been passing the last five months, I learn from the Proceedings of the Royal Geographical Society the sad news of Mr. Bates death. I feel the greatest grief for his loss, and I wish that not only his family, but also our Society, should rest assured that in all Brazil, and especially in the Amazons, his death will be deeply felt. Native of that vast and far-off province, as a child I learnt to admire the author of the Naturalist on the Amazons. Mr. Bates was one of the first to foretell the splendid future of the Amazonian valley, and to describe its bewildering splendours. Although his voyage to the Amazons was undertaken in his early manhood, time has not been able to efface his memory, and many an inhabitant of our regions still retains a vivid remembrance of the English naturalist. In 1889 I was fortunate enough to make his acquaintance, and great was my pleasure to find that he also remembered with delight his bold and perilous excursions in our forests, and recalled readily to mind episodes of that period of his life. On that occasion Mr. Bates offered me his portrait. I am convinced that the two provinces of Pará and the Amazons will make it a point of honour to place in their Congress Halls a large-sized reproduction of this photograph, as homage due alike to a modest savant and to English science.
> In conclusion, may I ask your lordship to be the interpreter of my sentiments on this occasion towards our Society and the family of the illustrious deceased?
> I have, etc,
> BARON DE SANTA ANNA NERY.
> Corresponding Member of the Royal Geographical Society.[157]

---

[156] There has always been a slight mystery about Bates's reaction to this award as he was painfully shy, for whatever reason, at acknowledging it in public. With the help of Francileila Jatene C Silva of the Museu Paraense Emilio Geoldi in Pará, I tracked down the original hand written record of the award in the private diary of Dom Pedro's visit to England. See also Appendix 2.

[157] Clodd, Edward, *The Naturalist on the River Amazons*, (London: John Murray, 1892): xxix.

The Brazilian Embassy in London confirmed they had no information of the whereabouts of these portraits and on my visit to Belém, I was unable to locate them.

Conclusion

Bates became a brilliant and efficient facilitator, who skillfully helped the council to develop a clear set of objectives and to achieve them. He was careful not to identify himself too strongly with any particular position – precisely the role of the good administrator or civil servant. He was often able to create a consensus out of previous disagreement and, having established a firm base, proceed to further action. Bates appreciated the importance of unspectacular essentials - following good meeting practices, timekeeping, keeping to an agreed agenda, and maintaining clear records. He listened carefully, paraphrased and sharpened the arguments of others. He drew people out, balanced conversations and made space for more reticent council members. If a consensus could not be reached, Bates at least helped the Council to understand the options before it. Watching over the activities of the Society at all levels, he gained universal respect. In short, his job was to support everyone to do his or her best. Bates encouraged full participation, promoted mutual understanding, and cultivated shared responsibility. Bates helped people to understand the Society's objectives and this enabled the Council to seek inclusive solutions, build sustainable agreements, and, as a result, grow in stature.

Recognition

As Douglas Freshfield said when addressing the Society after Bates's death:

> "There are three men among the Society's officers, who more than others, as I believe, built up its prosperity, (and secured for geography its proper place in education). These were Sir Roderick Murchison, Clements Markham, and Henry Walter Bates."[158]

[158] Royal Geographical Society Archives, report of the Evening Meetings, NS 14 182, 13.

Plate 16.
Bates's grave at plot G7/9 in East Finchley (Westminster Cemetary) London

Chapter 10

Bates in Perspective in Darwin's Century

A change in social and intellectual authority in England in the middle of the nineteenth century heralded the expansion in the numbers of scientists and the widespread dispersion of scientific ideas both at popular level and within education. As rival voices and factions sought to capture the prized role of becoming recognised as elite, it became clear that what was at stake was not only the authority of science in Victorian culture. A new public image of science was also being shaped. T H Huxley in particular actively engaged in formulating codes of ethics, strengthening professional organisations, and establishing professional schools. Educational institutions, public lectures, popular journals and newspapers, and books popularising science were crucial to this goal. The search for professionalisation was rife with conflicts among those who were within the scientific community but who had different ideological goals. Those deemed outside the emerging scientific community; mainly amateurs, posed another set of definitional problems. The notion of a working-class science further complicated an already difficult situation. So where did Bates fit in?

By the early 1860s, widespread recognition of Bates's accomplishments as a naturalist rendered him no longer an amateur. Nevertheless, was Bates a member of the emerging professional scientific elite? Although he corresponded extensively with his scientific peers especially with Wallace and Darwin eventually held posts in several amateur scientific societies, and was author of highly influential papers in the most prestigious professional scientific journals, Bates remained at the fringes of the elite. Moreover, professionalisation was as much about social status as about science and Bates seemed curiously indifferent to all this. Others like Huxley were social climbers letting nothing stand in their way, whilst men like Lyell did not need to do any climbing since they already possessed recognised status. Bates had no such ambitions and although he was fervent about his science, he seemed content to remain a social outsider, regarded as not especially crucial to the goals of the architects of emerging professional science.

However, Bates was not without career ambitions. He established a network of relationships with leading scientists during his first years back in England but as he remained outside their inner circles and could not obtain employment in the sciences, he was of necessity forced to take a job as an administrator at the Royal Geographical Society. In fact, because Bates spent the crucial decade of the 1850s away from London, mostly in the tropics where there were fewer social mores, he should have been able to establish himself more readily as a scientist of repute and not merely a gifted amateur naturalist. The boundaries of class and status, so sharply defined in England, were far less obvious in tropical outposts like Brazil, but it was not to be. His new environment with the Royal Geographical Society meant that he moved into a different circle, one that was focussed on Empire rather than on science, still an elite, but one that more readily accepted the amateur as the origins of the society were essentially explorer based, and a hotchpotch socially rather than elite. Once established he was able to make available a formidable administrative ability to the Society which they were not slow to value.

As a Unitarian, Bates made their dissenting voice the basis of his belief in nature. In the industrial heartlands of Leicester, the Chapel elite to which he belonged had developed a spirituality deliberately at odds with the fundamentals of Anglican power. They denied that all was subject to Divine aristocratic right and they refuted the Bishops' authority. The institutions of power, the universities, the law, and medicine were all in Anglican hands, but these new thinkers from an industrial background formed a rival forum in which their excluded, marginalised followers found the democratic republic of science appealing. They challenged the right of the nation's evangelical backbone to govern all things and decided they would not be blind to religious injustice. Bates's early years in stocking weaving Leicester, reading Southwood Smiths' reforming Unitarian bible, *The Divine Government*, cast a new light on the development of his scientific belief and religious non-belief. That, ultimately, is what makes Henry Walter Bates so fascinating. Like all supporters of Darwinian Theory, he helped shape our vision and while closing a window onto future immortality another was opened onto our prehistoric past.

This attempt to chart the life and work of Bates has involved consideration of his family and upbringing, his travels and explorations, his research work, his administration of the Royal Geographic Society, London, and his innate abilities and character. He enjoyed a reasonably long and diverse life with the central theme being his love for entomology. This final chapter examines his achievements as an entomologist in relation to the growth of our knowledge of this subject since the nineteenth century. It questions if his influence is still felt in these circles and, if so, how far it has been beneficial.

Bates's Position in the History of Science

I had many discussions with entomologists in Brazil that reinforced my opinion of Bates as a man of science. We concluded that, as Bates had no systematic training in biology this was probably crucial to his discoveries, as he had not been indoctrinated with the nineteenth century dogma that the differences found within species (intra-specific-variation) were unimportant and only the differences distinguishing species (inter-specific-variation) were of significance. With no professor to tell him what he should see in the specimens he found in the Amazons, he simply recorded what was actually there, and sometimes it was as novel as it was incontrovertible. Bates's greatest virtue was simply that he wrote as he saw, seldom taking the easy option of over-egging his adventures for the sake of a good story. He encountered primitive and perplexing people, but avoided the pitfall of caricaturing them as either noble Indians or idiots. For his time, he was the archetypal Englishman abroad, coping with the best and the worst of the world as he found it, and describing it truthfully.

As an early scientific invader of a relatively unknown land, Bates was largely isolated from European company. His interests had to range widely and the whole of natural history was in his province. His agent expected him to be a polymath, so Bates took note of rocks, plants, animals, the indigenous humans and everything around him with equal enthusiasm. Similarly, with extraordinary energy, each day's many findings were written down devotedly and comprehensively. There may have been setbacks, the climate was often hostile, the situation uncongenial, and the writer himself prey to debilitating illnesses, but the daily writings in his diary were usually optimistic. Men like Bates and Wallace were remarkable, leading strange and lonely lives as they made discoveries new to science, but this must have helped them during the long arduous years of absence from home.

Today's situation is different. Time always appears to be in short supply and it seems that no one can embrace all the scientific disciplines any more. Even if desired, and even if competent in various fields, the current system does not favour such widespread enthusiasm. An ecologist can and should take a wide view but the natural scientist today tends to specialise in certain species or genera through personal choice or necessity. Today no one can be called a naturalist in the old sense that Bates and Wallace were.

Entomology, the branch of science concerning the insects, is a vast subject and it was impossible for even a polymath like Bates to be a complete specialist in every branch of what was then known and understood. Entomology is dynamic not static and its development has evolved through the steady accumulation of minute facts. The modification of theories is largely due to important discoveries such as mimicry. The development of entomology has been a continuous process with occasional quantum leaps. Bates's own contributions to the advance of entomology were in four main fields: exploring the subject; collecting, taxonomy, and biogeography, the latter being the science of the distribution of species across space and time. He was also a competent administrator although this was mostly associated with the Royal Geographic Society rather than the Royal Entomological Society.

Bates was a distinguished scientist whose researches in entomology made an important contribution to the evolution of Darwinian ideas. His work as a taxonomist, one who studied the principles of classification and systematics, consistently aimed at a uniform standard of generic and specific definition and an adequate description of all creatures with a close similarity in appearance. His findings were based on considerable morphological (the study of shape, form and appearance) insight. His most important taxonomic contributions were in the realm of Amazonian butterflies, but in the earlier part of his life he did a considerable amount of work on beetles, which were his first entomological passion.

Systematic biology is the science by which scientists discover, describe, name and classify living and fossil organisms, and uncover their evolutionary relationships. It is central to the life sciences and, by devising a single agreed system of scientific names, it enables people around the world to communicate with each other about the diversity of past and present life on Earth. It provides guides for the identification of species and classifications that allow predictions to be made about their properties. The knowledge that is derived from systematics meets our basic need to discover and understand the world around us. By revealing the origins and evolution of life on Earth, we can begin to understand the history of our planet and our place within it. However, this knowledge is also essential for our well-being and long term survival, because organisms, including our own species, interact together in a complex web of life.

Bates made a broad generalisation that species were variable, often extremely variable, reaching this conclusion from his observations in the field well before the publication of any theory of natural selection, and indeed independently of other evolutionary views. He also recognised that, rather than being an immediate response to environmental differences, variations of particular taxonomic importance were inherited. It is true that the variations within a species were often more numerous and morphologically more marked to Bates than they are to us. This was because he was surrounded by vast quantities of living data and was able to focus on this alone. Conversely, much larger collections and more intensive study by improved methods now confirm that a great amount of variation of a mutational nature within a sexually reproducing species, however narrowly defined, is the rule. It is, indeed, much greater than is indicated in most descriptive faunas. As early as 1853, Bates summarised his findings in the axiom that species vary much more than is generally admitted to be the case, a theme he returns to frequently in his writings.

The basic concept of classification revolves around the species as the only categorisation having an objective definition. Species are defined as those individuals that can produce offspring freely with one another; therefore, what separates species is a reproductive barrier. Members of different species cannot interbreed in order to sustain a viable offspring able to breed in its own right no matter how similar they may appear to be. If hybrids do occur accidentally in the wild state, they are always infertile. All other units in classification, or taxons (hence taxonomy), are constructs and delineating them always involves some degree of unpredictability. The taxon that is above the species in the empirical order is the genus. A genus consists of a number of species that resemble each other in a specific way, which in butterflies is in the male genitalia, for if there is to be a natural barrier to interbreeding this is where it would be most effective. My studies in entomology show that even minor morphological differences between closely related species prevent mating, therefore a large white butterfly *Pieris brassicae*, cannot naturally mate with a small white butterfly, *Pieris rapae*, or with the green veined white butterfly, *Pieris napi,* even though they are closely related and belong to the same genus. The barrier to mating is a physical one in that the male genitalia will not connect with the female genitalia unless it is an exact fit and that only occurs within the same species. Marginal differences in the genitalia of different species within the same genera therefore function to screen out the species that are generally incompatible by the use of these mechanical preventive measures.

A number of related genera form a family and a number of these form an order, related orders a class, related classes a phylum and related phylum, a kingdom. Bates was aware of all this but, as intrinsic issues such as genes had not yet been discovered or described, he could only classify his specimens through their physical characteristics. A number of physiological features can determine classification up to the hierarchy of families, which were the basics of his revelations in describing species that appeared to be similar but were in fact quite different. This raised the question why should they appear to be so alike yet so diverse, and this led to the idea of mimicry.

Bates was not the first to recognise variants within a species. Linnaeus accepted this for a number of species, but he was not able to investigate in detail what was meant by intra specific variation (relating to variation within a species). It is only within recent years that biologists have been able to do this, and methods, terminology, collection of data and correlation of results still have far to go before valid generalisations can be made. Variation is different in kind and degree in diverse groups of insects and it may have singular meanings to various investigators. What Bates did was to call attention to its widespread occurrence in insects, notably in his papers read at the Linnean Society. Many aspects of intra specific variation have since been intensively studied and the results confirm and greatly extend Bates's original conclusions.

I also believe that Bates contributed a great deal to descriptive taxonomy in preparing and publishing, always with accurate illustrations, accounts and revisions of species and genera new to science. He also did much to increase and maintain a high standard of clarity, uniformity and conciseness in his descriptions of species; which are considered today the finest examples available of such work. Excellent pieces are the generic descriptions in papers published by the Linnean Society; e.g. Descriptions of Fifty-two New Species of *Phasmidae* from the Collection of Mr. W. Wilson Saunders, with Remarks on the Family. By Henry Walter Bates, Esq. 1865.

Any reader with little or no understanding of the Phasmids, could understand Bates's portrayal of *Bacteria cyrtocnemis* in this paper:

> "Head with the crown much raised, the front part of the elevation having two strong and short, acute spines directed obliquely forwards. Thorax unarmed. Abdomen cylindrical to the apex; the apical segment equal in length to the preceding, its apical edge triangularly emarginated; the vaginal operculum narrow, not at all convex, pointed at the tip, and not passing the apex of the abdomen; anal styles short, straight, obtusely pointed. Legs moderately stout, scored between the raised lines, which are strongly elevated and compressed; the fore and hind legs are elongated and simple; the middle femora are strongly curved, and have, at one-fourth their

> length from the base, on each side, a flattened expansion divided into two tooth-like points. The basal joint of all the tarsi is much elongated. The colour of the entire insect is pale green, and in life it bears the closest resemblance to a stalk of grass, or rather the midrib of a palm-leaflet. Hab. One example, taken at Ega on the Upper Amazons."[159]

He is even more eloquent when describing butterfly behaviour as he does in '*Contributions to an Insect Fauna of the Amazon Valley*', a paper read before the *Linnean Society*, on 21st November 1861, in this case referring to the flight habits of both the Heliconias and the Ithomines. The mode of flight of the members of the two groups is somewhat different.

The Heliconii and Eueides move along in a sailing manner, often circling round for a considerable time, with their wings horizontally extended. The species of the Danaoid group, for the most part, keep near the ground, and have a very slow irregular flight, settling frequently. They are all of social or gregarious habits. Not only do individuals of the same species congregate in masses, but also the set of closely allied species, which people a district, keep together in one or more compact flocks. I noticed in four districts rich in *Danaoid Heliconiidae*, where I collected, that about half the species of Ithomia flew together in one circumscribed area of the forest, and the other half in a second similar locality, the rest of the tolerably uniform wooded country, in each case, being nearly untenanted by them.

> "The larger species (Heliconii, Lycoreae) -frequent flowers, probing the nectaries with their proboscides; but the smaller kinds (Ithomiae), and the members of the Danaoid group generally, are very rarely found thus occupied: I noticed them sometimes imbibing drops of moisture from leaves and twigs. The fine showy Heliconii often assemble in small parties, or by twos and threes, apparently to sport together or perform a kind of dance. I believe the parties are composed chiefly of males. The sport begins generally between a single pair: they advance, retire, glide right and left in face of each other, wheel round to a considerable distance, again approach, and so on: a third joins in, then a fourth, or more. They never touch: when too many are congregated, a general flutter takes place, and they all fly

[159] Descriptions of Fifty-two New Species of Phasmidae from the Collection of Mr. W. Wilson Saunders, with Remarks on the Family. By Henry Walter Bates, Esq. 1865. *Transactions of the Linnaean Society;* Vol xxv, (1865) 321-359 with two plates.

off, to fall in again by pairs shortly afterwards. The species that I have seen most frequently employed in this way is the *Heliconius rhea*, a glossy blue-black species, with bright yellow belts across its wings."

I have been to the forest and witnessed such events, but the reader who has not seen them in life is left in no doubt by Bates's description of the appearance of these butterflies.

Bates was not alone in the study of mimicry. Wallace wrote about this phenomenon too, but it is impossible now to separate the different part played by either of them in the general scheme of the theory. Bates did more than Wallace in the actual describing of families and genera and in arranging the latter, but equal responsibility rests on both for the choice of basic principles and for the sequence of families. It was, however, Bates who was the first to publish his ideas. The great advance made in the taxonomy of butterflies by Bates was due to the care he took to prepare descriptions of genera and families on a uniform basis from the actual examination of specimens. Bates's descriptions are models of accuracy, and their originality raised the whole work to a peak of excellence that has never been exceeded by any subsequent writings.

However, it was not only with butterflies that Bates contributed to entomological science. His work as a coleopterist (a beetle specialist), is embodied in Godman and Salvin's work, *Biologia Centrali Americana.* Of the first part of his contribution, which deals with two large families, the *Cicindelidae,* or tiger beetles, which stand at the head of the whole order, and the *Carabidae,* or ground beetles, a reviewer in *Nature* (26th November 1885) wrote as follows:

"Its author has long been known as an entomological systematist, for it is now nearly twenty-five years since he inaugurated a rational classification of the Rhopalocera, or butterflies. He has been recognised, since the death of Baron Chaudoir, as the one entomologist possessing an extensive yet intimate knowledge of the Carabidae of the whole world. But Chaudoir died without leaving behind him any general work on the classification of that family. It is therefore a matter for congratulation that the author of this beautiful volume has presented us with a systematic arrangement as complete as the faunistic nature of the work permitted. The family Carabidae is of such enormous extent, 12,000 species being known, with a vast number of others to come, that the necessity of some series of

> intelligible aggregates, subordinate to the division, but superior to the tribe or sub-family, is undeniable, and Mr. Bates attempt to furnish such a series is therefore of great value."

In two papers printed in the *Transactions of the Entomological Society* in 1890 and 1891, the additions to the *Carabidae* of the Mexican fauna since the publication of the first part of the *Biologia* were described by Bates and updated. At the time of his death, Bates was engaged on an improved classification of that family, but the true value of his work was not in introducing the use of new characters but in selecting what was best in the various older systems. While doing this he was able, with his wide knowledge, to settle the proper positions of a number of new forms whose affinities up to then had been doubtful.

Whilst the great contributions of Bates to Amazonian entomology are acknowledged, it may be asked why, having accepted the general theory of evolution and the Darwinian model of natural selection, albeit with some reservations, Bates did not in due course attempt to work out a classification of insects on evolutionary principles. There are several possible answers to this question.

First, there was his clear principle that classification was essentially a practical matter: it was to enable entomologists to identify the insects they had before them.

Secondly, this seemed best achieved by assembling in one group or in closely associated groups, insects with a maximum number of common characteristics.

Thirdly, we have to recall Bates's natural caution in theoretical matters. He did not like speculation beyond a firm basis of fact. It may well be that he was sometimes too cautious and that he would have made more valuable additions to theoretical taxonomy had he been less so.

Fourthly, he knew that the course of evolution in insects is complicated and, as far as we can at present ascertain, not simply linear. It is frequently termed reticulate, although the reticulation is itself not uniform in nature or cause.

Fifthly, Bates knew better than many entomologists of the time that, in order to determine a reliable phylogeny, a reasonably complete fossil record was essential and this was not available for insects.

Sixthly, he may well have considered that a natural classification in the Linnaean sense was essentially phylogenetic and there was no need to look into the matter further. In theory, acceptance of evolution surely means that phylogenies can be traced, given the evidence. Many entomologists think that with the exception of a few major butterfly groups and some minor assemblages, reliable evidence is not yet available. Hence, it is argued, most proposed phylogenies are extremely doubtful approximations as to what happened in the course of evolution. Phylogenists are often sceptical about schemes other than their own. It can, however, be said that a phylogenetic scheme, when properly explained and with evidence clearly stated, is useful in that it summarises the application of known facts and suggests lines for future investigation.

Modern developments in entomology throw some light on problems of evolution and heredity in ways not known to Bates. How far any of them, or a synthesis of them, may lead to improvements in taxonomy and to a sound phylogeny of insects, and how far such phylogeny can be made the basis for a practical taxonomy, remains to be seen.

It would be absurd to criticise Bates because he did not know about chromosomes, DNA, growth hormones and other topics that are now familiar to most entomologists and even to non-specialists. What Bates in his time did for taxonomy was to continue its growth along known and sound lines, greatly advancing knowledge of hitherto unknown species and preparing the way for future intensive studies.

Bates however had certain advantages as a butterfly hunter in the 1850s. He knew what is now sometimes forgotten, that taxonomy must be the basis for the study of insect ranges and distribution. The units (taxa) dealt with must be defined consistently before their ranges or habitats can be mapped or tabulated. Again, as a classifier, Bates knew the importance of giving due weight to all species and genera as evidence for any conclusions reached. There is always a danger of missing the truth by subjective bias if the ranges of a few species or genera or even families are picked out for special consideration. This can also be the case if their ranges are emphasised in support of this or that hypothesis, unless at the same time an analysis of the known fauna is provided. When reaching conclusions on a bio-geographical issue, Bates presented all the known evidence, positive and negative, fairly and lucidly to the limits of data and knowledge then available.

We have to remember that Bates's studies of Brazilian butterflies and especially of mimicry were for the most part pioneer studies, in that they were concerned with areas from where insects were not readily available and at a time when insect classification was imperfectly known. An enormous amount of research in entomology has been carried out since

Bates's investigations, but it was his early work associated with his travels in the Amazons that was the foundation for many of these future studies.

Bates was always an acknowledged Darwinist, but his views on evolution were closer to Wallace's than to most of Darwin's other supporters. In contemporary terms, Bates was a *group selectionist*; that is someone who believed that infighting was harmful to society and that group behaviour was the key to evolution. Darwin, on the other hand was an *individual selectionist*, someone who believed in the inevitability of gain based on the principle of individual selection, and integral to that was the necessity for internal strife. The meanings of these two different approaches to Darwinism are fundamental and have far-reaching implications in human affairs. Darwin was convinced that in the non-human world selection acts at the level of the individual. Bates and Wallace, on the other hand, thought that the limiting factor was group selection. It seems sad to me that Darwin wrote to Wallace. 'We shall, I greatly fear, never agree on this'. Wallace remained adamant in his belief that human morality was subject to different laws and could not be explained simply by evolutionary theory and Bates always supported him in this.

The Russian revolutionist Peter Kropotkin draws attention to this view of Bates in his book, *Memoirs of a Revolutionist* published in 1899 in which he says:

> "I found in a lecture of a Russian zoologist, Professor Kessler, a true expression of the law of struggle for life. 'Mutual aid', he said in that lecture, 'is as much a law of nature as mutual struggle; but for the progressive evolution of the species the former is far more important than the latter'. These few words...contained for me the key of the whole problem. When Huxley published in 1888 his atrocious article, 'The Struggle for Existence: a Program', I decided to put in a readable form my objections to his way of understanding the struggle for life, among animals as well as among men, the materials for which I had accumulated during a couple of years. I spoke of it to my friends. However, I found that the comprehension of 'struggle for life' in the sense of a war-cry of 'Woe to the weak', raised to the height of a commandment of nature revealed by science, was so deeply in rooted in this country that it had become almost a matter of religion. Two persons only supported me in my revolt against this misinterpretation of the facts of nature. The editor of the Nineteenth Century, Mr James Knowles, with his admirable

> perspicacity, at once seized the gist of the matter, and with a truly youthful energy encouraged me to take it in hand. The other was Bates, whom Darwin has described in his autobiography as one of the most intelligent men whom he ever met. He was secretary of the Geographical Society, and I knew him. When I spoke to him of my intention he was delighted with it. 'Yes, most assuredly write it,' he said. 'That is true Darwinism. It is a shame to think of what "they" have made of Darwin's ideas. Write it, and when you have published it, I will write you a letter in that sense which you may publish.' I could not have had better encouragement, and began the work which was published in the Nineteenth Century under the titles of 'Mutual Aid among Animals', 'among Savages', 'among Barbarians', 'in the Mediaeval City' and 'among Ourselves'. Unfortunately I neglected to submit to Bates the first two articles of this series, dealing with animals, which were published during his lifetime; I hope[d] to be soon ready with the second part of the work, 'Mutual Aid among Men', but it took me several years before I completed it, and in the meantime Bates was no more among us."

One has to remember that Kropotkin was onetime secretary of the Russian Geographical Society and therefore a working contemporary of Bates.

Throughout his life, Bates was a staunch Darwinian but he was not an instigator of the idea of natural selection or of the theory of evolution. However, he was a player from the beginning when colleagues Darwin and Wallace set out their belief in evolution and natural selection.

The essence of these thoughts was that many more individuals of a species are born than can normally live to maturity and breed successfully so inevitably there is a struggle for existence. There is also constant individual variation in innumerable characteristics within a species, some of which may affect a species' ability to survive and reproduce. The successful parents of a given generation may therefore differ from the population as a whole. Finally, there is likely to be a hereditary component to a great deal of this variation, so that the characteristics of the offspring of the successful parents differ from the distinctiveness of the previous generation, and are more similar to their parents.

If this process continues from generation to generation there is a gradual transformation of the population, and the frequencies of characteristics associated with better survival ability or reproductive accomplishment increase over time. These changed characteristics originate by mutation, but mutations affecting a particular trait arise all the time regardless of whether or not they are favoured by selection. However, evolution or Darwinism has also itself evolved. Modern thinking about Darwinism suggests that goodwill and collaboration are as much part of the human condition as ill will and competition, something that was a puzzle to Darwin, Wallace and to a lesser extent Bates.

It was not Darwin, Wallace or Bates but a contemporary, Herbert Spencer (1820-1903), who coined the phrase, *survival of the fittest.* When Darwin's *On the Origin of Species*, was published, Spencer quickly saw the parallel with natural selection and transferred his mind to the process of evolution, becoming a staunch social Darwinist. In due course capitalists took what they thought were the lessons of Darwin's book and applied them to human society. Their conclusion was that people got what they deserved, but the decisive factor was genetic, rather than moral. The fittest not only survived, but also prospered. Social Darwinists thought that measures to help the poor were a waste of time, since the poor were unworthy of the effort and, as they were not among the fittest, would in any case disappear.

Spencer was one of the first Londoners Wallace sought out on his return from the Malay Archipelago. He was accompanied by Bates who by then had read and 'been immensely impressed' with *First Principles.* Bates and Wallace hoped for enlightenment regarding the great-unsolved problem - the origin of life itself. They realised that *On the Origin of Species* had deliberately left that problem 'in as much obscurity as ever'. Both he and Bates eagerly looked to Spencer as the one man living who could give us some clue to it.

Wallace's account of that first meeting says much about their desire to probe deeply the fundamental questions posed by evolutionary theory. They were impressed that Spencer spoke appreciatively of what they had both done during their travels for the practical explanation of evolution, and hoped they would continue to work contentedly at minor problems in the future. (How condescending of him!) Spencer said it was too fundamental a problem to solve at present. Wallace in particular was less than impressed with this, as he considered his own developing evolutionary philosophy would enable him to approach questions about the origin of life and the course of human evolution by epistemological avenues that Spencer and Darwin neglected or ignored.

People certainly compete, but they collaborate too and they also have compassion for the underprivileged and frequently try to help them. This was the type of behaviour for which Darwin had no explanation. Exactly how humanity became human is still a matter of debate. However, there are some well-formed hypotheses that rely not only on the idea of individual competition, but also on social interaction which may be confrontational, intermittently bloody, but often collaborative.

Progress in modern Darwinism is marked by the identification of the central role of trust in human evolution and by the fact that people who are related collaborate nepotistically. It takes extreme provocation for someone to condemn a relative with shared genes. Trust allows the unrelated to collaborate too, by keeping score of who does what and when and rewarding them accordingly, as well as punishing those who cheat. The human mind is able to identify a large number of individuals, manage its relations with them, detecting the dishonest or greedy, and seeking retribution for misdemeanours, sometimes even at some cost to itself.

The new social Darwinists, those who view society rather than the jungle as the 'natural' environment in which humanity is evolving and to which natural selection responds, see the ranking by wealth of which Spencer so approved as one example of a wider tendency for people to try to outdo one another. Thus, to make an economy work, both collaboration and competition are needed. Today, Darwinism continues its own struggle for survival. If its ideas are right, the first stirrings of life that evolution shaped into humanity will rarely stray too far off course. That is where Bates re-emerges, as an early protagonist of the idea that evolution was not just about the species and the pressures of its environment, but as much about the interactions that take place within a society. These were the thoughts that contributed to the theory of mimicry.

In the end, it was the aura of the Amazons that created Bates's fame, and his obituaries unanimously recalled this defining episode of his life and the sobriquet, *Bates of the Amazons*, by which he was affectionately known.

In an extensive obituary in *The Fortnightly Review*, the novelist Grant Allen recalled an evening at Edward Clodd's North London home towards the end of Bates's life when Bates talked of his travels on the Amazons. Among those attending were the writer, Samuel Butler and the African explorer, Paul du Chaillu. Bates reminisced about the years he had spent on the river. Allen described Bates as an old man whispering about how he had at times almost starved to death, of how he had worked with slaves like a slave for his daily rations of coarse food and how he had often faced perils more appalling than death. There were times, he went on, when he had risked and sometimes lost everything he possessed on earth. The emotion with which he talked of these events brought tears to the eyes of the grown men who heard him.

When he finished one of those present was heard to say, 'Oh, if we had only had a phonograph to take that all down, accent, intonation, and everything-exactly as he spoke it.'[160]

This is a complex image by which to remember Bates with men of science and literature together at his feet listening to the tales of his great adventure. We can only guess at the Amazon Bates describes for it is no longer there as he saw it. Today we can only speculate, but we do know the contours, limitless nature, incredible hardship, broken health, intimate comradeship and freedom that were all part of his extraordinary journey.

When engaging in the history of this single event in Bates's life it is impossible to ignore the politics of class and race that dogged his ambitions after his return from Brazil. Why was it that his peers, men of substance, were compelled to draw attention in the midst of all this fascination to the traveller's accent, his tone, his provincial origins, his lack of sophistication? It is never actually said in so many words by any of them, but it is the illiterate wife and the ordinary family that appear to define the man. It is an appalling anticlimax. The scientific elite at a moment of ascendancy behaved egregiously, confirming the foolish way men can sometimes act towards each other. It was for these reasons that Bates, the consummate natural scientist, the finest entomologist of his day, was shut out of the inner circles of the nineteenth century's world of science, and that is why, with the exception of a scientific minority, he is now largely forgotten. However, to the scientific minority Bates had distinguished himself during his long, hard, and ambitious life as a principal player in the new science of evolution. He was one of the foremost insect collectors of his time and a first-rate taxidermist and taxonomist. At the time of his death in 1892, the hosier's son from Leicester, so influenced by Victorian values and nonconformist morality, had earned the epitaph the 'Butterfly Hunter'.

One can still visit the manuscript collection in the Entomology Library of the British Museum of Natural History in South Kensington and see several of the notebooks Bates kept while on the Amazon. They are beautiful, simple exercise books filled with delicate watercolours of butterflies and beetles and of such clarity that they seem to have been

[160] Clodd, Edward, *The Naturalist on the River Amazons*, (London: John Murray, 1892) xxix.

contemporarily painted. In a careful, precise hand, Bates has catalogued his collecting and there for all to see, he says '*some mistake here.... I think I have ticketed the wrong specimen; the insect is not Pleuracanthus*'. This, in its meticulous attention to detail, is the way to remember Henry Walter Bates.

Chapter 11

Summing Up

Without doubt, employment at the Geographical Society kept Bates away from his scientific work. His salary at the Royal Geographical Society was eventually £300 per annum; in relative terms not a bad salary in the 1860-90s. However, his employment demanded a professional lifestyle that cost a good deal to maintain. With this and the obligations of a growing family, Bates had to supplement his salary by spending a great deal of his spare time on hackwork. For many years, both he and Wallace supplemented their incomes by marking geography examination papers. Bates also undertook editing and ghost writing for John Murray, who paid well for such work. As for many men from a frugal beginning, with careful management better days followed. This was so for Bates as he eventually inherited and earned enough money to buy his London house in Tufnell Park outright as well as his summer home in Folkestone. He was then able to educate his children, marry off his daughters, and set up two sons as sheep farmers in New Zealand and another as an electrical engineer.

However, throughout his life after the Amazon he was dogged by ill health, partly aggravated because he always smoked pipe tobacco. He eventually suffered from chronic pulmonary disease, which finally killed him. As his workload increased and his health worsened, his scientific work suffered proportionately. Darwin however relied on him for data and entomological specifics to the end of his life. Before securing his position at the Royal Geographical Society in 1864, Bates was forced by economic needs to sell his private collection of butterflies. From then on he concentrated on the *Carabidae*, the ground beetles that had first caught his eye when he was a boy, and on which he eventually became recognised as the world authority.

After the temporarily elevated position he enjoyed with Wallace and Darwin in the area of evolutionary theory, from 1870 onwards he concentrated his scientific activities on the cabinetwork he deplored; the process of naming, arranging, and labeling specimens which could have been done by a man of lesser ability.

Ironically, it was after Bates had abandoned the effort of making further independent theoretical contributions to natural history that he began to receive recognition from the scientific establishment. For the last twenty years of his life he was a familiar and influential figure in Victorian scientific circles, to be seen regularly at British Association gatherings and at the evening meetings of the Entomological and Linnean Societies. He was also frequently present among his fellow travellers at the Kosmos Club. Undoubtedly as a systematic naturalist, competent explorer and enlightened geographer, he did a great deal of good in bringing men together whose interests in natural history and travel complemented each other, and he was celebrated as the author of *The Naturalist on the River Amazons*. However, one is still at a loss to determine why *The Naturalist on the River Amazons* was the only book he ever wrote. This should not have been caused by poor health; Darwin was also unwell when he wrote his most important books, but then he did not have to work as hard as Bates did. Nor can financial anxiety be an excuse; Bates was in fact quite well off for his time and Wallace, after his return from the Malay Archipelago, was much less secure yet wrote prolifically. Nor is there any evidence that the litheness of his mind or his breadth of interest diminished after Brazil. Evidence shows that he remained a stimulating conversationalist for the rest of his life and people liked his company: As Clodd remarked after one of Bates' Sunday evening gatherings:

> "There was a wonderful freshness in all that he said and a wonderful charm in the way he said it. His sentences were broken by curious hyphen like pauses. But how perfect they were in construction; clear-cut, pure English, so that, taken down not a word need have been altered or transposed. Never did the listener leave without being the richer for some fruitful idea, some fresh aspect of familiar things, evidencing the power of the speaker in seizing upon the relation of a particular fact or theory to the totality of knowledge."

For the first four years after his return from the Amazons, the motivation persisted, but when he finally decided not to travel again, and gave hostage to fortune by marrying, the impulses born on the river of the sun diminished and lost their power to motivate him.

He remains an enigma and we must consider, as our only explanation for what happened, that the experience of travel had, for Bates, more hidden significance than the pursuit of science would ever hold for him. Bates remained permanently affected by his Amazon experience as men are by their passions. The period on the river was in fact the central happening of his life, from which all his original scientific work emerged.

Bates and Sub Personalities

Perhaps the most traumatic event in Bates' life was leaving the Amazons and adjusting to his return to England. In the face of distress or great disappointment, as this event must have been, it is not unusual for human beings to create a sub-personality, becoming some other person playing out a new role or part. As a means of dealing with an otherwise unbearable event, Bates may have established a new personality for himself, completely divorced in his mind from the person he had been in Brazil. There would then have been two Bates personalities; one the lone collector in the forests, the other post Brazil. The development of a sub-personality is an extension of normal, common behaviour. All of us modify our personality depending on our situation. The significance of dual personalities is that the individual is not aware of the change, so the personality Bates took on in order to cope with change would not have needed any connection with his normal personality. Various means are used to reinforce sub-personalities; workaholic tendencies being one of the obvious ones, getting married another, reaching the optimum in achievements or self-actualising, yet another.

Conceivably this was the driving force behind his time at the Royal Geographical Society where he was the able administrator and sage, leaving behind a legacy of good judgment given without fear or favour to anyone who asked. In the Society, he fulfilled a role that was as significant as his earlier scientific days and, in retrospect, much more fundamental and long lasting. He should be remembered for this as much as for his theory of mimicry. Without Bates in the assistant secretary's chair, the Society would probably not have been the driving force it was in establishing geography as an academic subject based on scientific principles. Bates was able to succeed brilliantly in one field of science, and then go on to excel, and possibly achieve even greater heights, in his work for the Royal Geographic Society. Perhaps this was the pinnacle of his achievements.

Appendix 1

Bates's butterflies

Bates's butterflies are important because they tell us more about the man and his meticulous habits. Any specimen to be of scientific value must bear a label giving the all-important information about its collection available at the time of capture. Bates's specimen labels, in his own hand, indicate the extent and detail of his precise journeys. Tracing Bates's original specimens is often difficult as it is the fate of most individual gifts to major museums to be assimilated within the core series. This enriches the overall collection as a foundation for taxonomic study. From the historical perspective, this often results in vital information being lost but individual collections can be reconstructed as long as specimens still bear their unique labels. Restoration work has recently been undertaken on Bates's Amazon butterflies now held in the BMNH. [161]

Some of his specimens are illustrated in Bates's, Contributions to an insect fauna of the Amazon valley. Lepidoptera: Heliconiinae. *Journal of the Proceedings of the Linnean Society of London.* (1862) vol XXIII plate 56.

Plate 17.

[161] Kim Goodyer and Phillip Ackery in *The Linnaean, Proceedings of the Linnaean Society of London*, (2002) vol 18, 27-59.

Figure numbers are top figs 4, 5, 6, second row figs 4a, 6a, third row figs 7, 8, fourth row figs 7a, 8a.

Key to the plate.

This plate illustrates various species of *Leptalis* (4, 6, 7, and 8) that mimic the *heliconid* butterflies with which they associate (4a, 6a, 7a, and 8a) the *Leptalis* thereby gaining protection by looking like the *heliconids*, which are probably toxic to predators.

Description of the plate.

Fig. 4. Leptalis *theonoë*, var. *leuconoë*. - São Paulo de Olivença.
Fig. 4a. *Ithomia llerdina* (Hewitson). - São Paulo de Olivença. This *Leptalis* appears at first sight a distinct species, but it is a modification whose adaptation is complete.
Fig. 5. *Leptalis nehemia* (Bates). - New Granada. Figured to show the normal form of the family *Pieridae*, to which *Leptalis* belongs. The contrast this species shows with the remainder of its family points to the conclusion that all other *Leptalis* illustrated divert from the family type after a long continued process of adaptation to look like the *Heliconiidae* as each species associates and flies with its *Heliconian* model.
Fig. 6. *Leptalis theonoë*, var. *argochloë*. - São Paulo de Olivença.
Fig. 6a. *Ithomia virginia* (Hewitson). - São Paulo de Olivença. The links of modification can be seen also with respect to this apparently distinct *Leptalis*.
Fig. 7. *Leptalis amphione*, var. *egaëna*. - Ega (Tefé).
Fig. 7a. *Mechanitis polymnia*, var. *egaënsis*. - Ega (Tefé).
Fig. 8. *Leptalis orise* (Boisduval). – Cuparí; also Cayenne.
Fig. 8a. *Methona psidii* (Linnaeus). - Cuparí; also Cayenne.

Bates documented this material, in a successive series of articles read at the Linnean and Entomological societies meetings under the title *Contributions to an Insect Fauna of the Amazon Valley:* 1860-61 covering *Papilionidae.*[162] 1862 *Nymphalidae-Danainae, Ithomiinae, Heliconiinae* (in part).[163] 1864 *Nymphalidae*

[162] Bates, H W, Contributions to an insect fauna of the Amazon valley. Lepidoptera *Papilionidae, Journal of Entomology* (1861), 1, 218-245.
[163] Bates, H W, Contributions to an insect fauna of the Amazon valley. Lepidoptera *Nymphalidae. Journal of Entomology* (1865), 2, 311-346.

introductory text and *Heliconiinae* (in part), *Nymphalinae, Biblidinae* (in part).[164] 1865 *Nymphalidae, Biblidinae* (in part), *Apaturinae, Charaxinae, Morphinae.*[165]

Bates's Descriptions

Apart from labels, specimens had to be described and Bates's vibrant descriptions of his butterflies bring alive the glorious living insect that would otherwise lie hidden behind its formal description. Contemporary writers often referred to Bates's work and used his descriptions in their own work, as was the case with the Rev J G Wood in his book *Foreign Insects* where he says:

> "...this extremely rare insect, which is figured here, is not in the British Museum. [So it must have been in private hands] Mr. Bates, who named it after his daughter, captured it."

He then quotes Bates's description of the Butterfly and its habits:

> "Slaty green, silky fore wings above, with many black or dusky variously shaped spots, nearly all of which are margined with a paler hue. Besides these dark spots, there are ten or twelve pale brown spots, one, or two between each of the longitudinal nervures. Margins of the wings black. Hind wings with a row of black eyes running parallel with the margin and edged with green-some of them have slate green pupils.
>
> Beneath, the forewing ochreous at base, the rest of the wing dark brown, with three belts of white spots. Hind wing, clear saffron yellow; outer margin black, with ochreous spots between."

[164] Bates, H W, Contributions to an insect fauna of the Amazon valley. Lepidoptera, *Nymphalidae*'. *Journal of Entomology* (1864), 2, 175-213.
[165] Bates, H W, Contributions to an insect fauna of the Amazon valley. Lepidoptera, *Nymphalidae*'. *Journal of Entomology* (1864), 2, 175-213.

FIG. 345.—Ageronia Alicia. Upper side. (Slaty green and black.)

Plate 18.

Ageronia Alicia **named after Alice Bates**

FIG. 346.—Ageronia Alicia. Under surface. (Slaty green and black.)

Plate 19.

Ageronia Alicia **named after Alice Bates**

> "This fine and large species was met with only at São Paulo de Olivença, Upper Amazons. It has the same habits as its congeners, but it is much swifter in flight. Although I saw several, I was able to capture only one example. The expanse of wing is three inches and three-quarters."[166]

Wood again uses Bates's descriptions for the genus Zeonia this time with details taken from a letter from Bates to Adam White, another contemporary collector, dated Tefé, 2nd May 1857, and reprinted for general interest in the transactions of the entomological society.[167]

> "The beautiful Zeonia, of which I sent you a large series last July, I met with in a part of the forest near Ega, which I had traversed and examined before, many times, in all seasons. The first, specimen I found was a straggler in a different part of the forest. On July the 21st, after a month of unusually dry and hot weather, in ascending a slope in the forest by a broad pathway mounting from a moist hollow, choked up with monstrous arums and other marsh plants I was delighted to see another of what had always been so exceedingly rare a group of butterflies; it crossed the path in a series of rapid jerks, and settled on a leaf close before me. Before I had secured it I saw another, and then shortly after a third. I mounted to the summit of the slope, followed a branch pathway which led along the brow of the ridge, without seeing any more, but returned again to examine well the exact spot where I had captured the three, for it very often happens that a species is confined to a few square yards of space in the vast forest, which to our perceptions offers no difference throughout its millions of acres to account for the preference. I entered the thicket from the pathway, and a few yards therein found a small sunny opening, where many of the Zeonia were flitting; about from one leaf to another, meeting one another, gamboling, and fighting; their blue transparent tinge, brilliant crimson patch, and long tails, all very visible in the momentary intervals between the jerks in their flight. I was very busy, you may imagine, at first in securing a supply of specimens; I caught perhaps 150, two-thirds of which fell to pieces in the bottom of the net, so fragile is their

[166] Rev J.G.Wood, *Insects Abroad*, (London: Longmans 1892) 605-606.
[167] Proceedings of natural history collectors in foreign countries. Mr. H. W. Bates. *The Zoologist* 11:3726 – 3729

texture. [So collecting perfect specimens for cabinets was not necessarily easy] I then paused to look around the locality, and endeavoured to find the larva; and pupae.

I walked through the thicket in all directions, and found the space peopled by the species was not more than from twenty to thirty square yards in extent: as far as the eye could reach, the leaves were peopled with them; it is possible the brood belonged to some one tree. The only two pupa; I could find it is true, were on two distinct kinds of trees, but this is no proof that the larvae may not have fed on one tree only. I was disappointed at not finding the larvae, although I searched well during this and the three following days.

On the second day, the butterflies were still coming out; on the third, they were much fewer, and nearly all worn; and on the fourth day, I did not see a single perfect specimen, and not a dozen altogether. During all the time I worked in the neighbourhood of the city of Pará I found but one specimen of a Zeonia. This was in 1848. The next time I saw the genus was at Alter do Chao, where I took a few of a very small long-tailed species at flowers. At Ega, a few miles up the Teffé, I took one of another very handsome species at flowers, very distinct from all the others.

The colours of this large genus are very similar throughout, and are generally black, scarlet, and white, the scarlet in some species giving way to yellow.

Zeonia faunus, has the greater part of both wings nearly transparent. The upper wings are surrounded with a band of black, powdered with tiny grey specks, and a rather jagged bar of the same line runs through the middle. The upper part of the lower wings is edged with jetty black and the lower part with black, powdered like the upper wings. In the midst of the black are two spots of scarlet, with a slight tint of orange, something like that lovely but too fugitive pure scarlet of the colour-makers, one of the spots being large and oval, and the other small and circular. The colours are nearly identical on the upper and under surface of the wing. There is only one specimen in the British Museum. [Indicating scarcity presumably.]

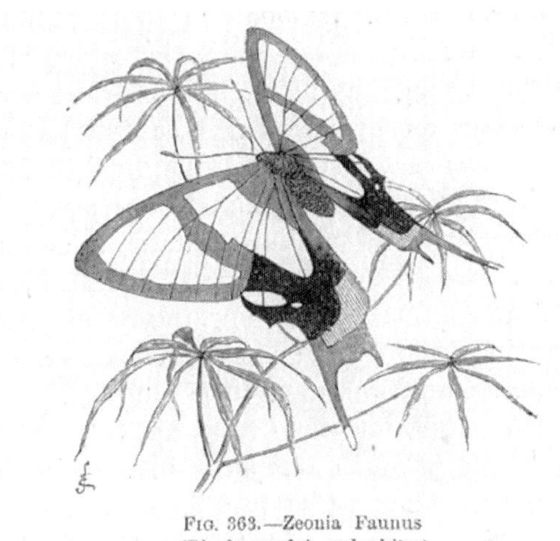

Fig. 363.—Zeonia Faunus
(Black, scarlet, and white.)

Plate 20.
Zeonia Faunus

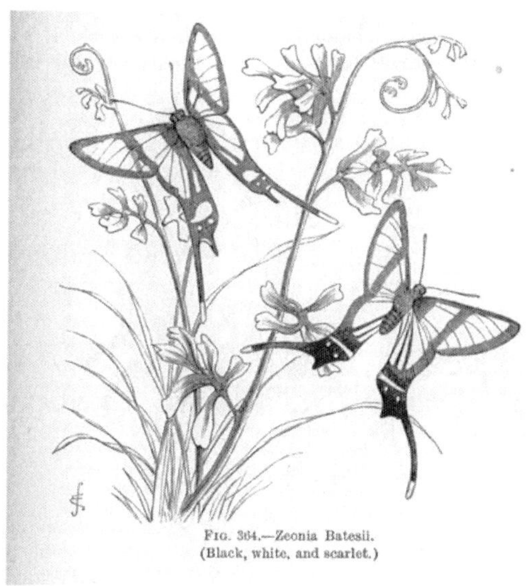

Fig. 364.—Zeonia Batesii.
(Black, white, and scarlet.)

Plate 21.
***Zeonia batesii* named after its discoverer, Bates**

In the (above) illustration are given the two sexes of *Zeonia Batesii*, in order to show the curious difference of shape as well as colour between the males and females. The colours are arranged in much the same manner as those of the preceding species, the edges of the wings being black, the centre transparent, and a scarlet patch on the lower wings, which in this species is followed by two little white spots. At the extremity of the tails of the lower wings, there is a short streak of yellowish white. The reader will probably have noticed that, whereas in the males the projections of the lower wings are tolerably straight, in the females they diverse considerably, and turn a bold curve. Both these insects are natives of the Amazons district."

Bates's collecting day appears to have been methodical. He writes from Tefé to his brother Frederick that starting early in the morning he donned a coloured shirt, a pair of trousers, a pair of common boots, and an old felt hat. Then he would sling his gun over his left shoulder, in one barrel a number 10 shot, the other a number 4. His net was in his right hand and his body bedecked with bags and thongs for carrying his specimens. In his pockets were the papers needed for wrapping small birds, wads of cotton, a box of powdered plaster, and a box with damp cork for micro-Lepidoptera. To his shirt he attached his pincushion with six sizes of entomological pins. This artifact is now preserved in the Royal Geographical Society archives. Attired in this manner he would wander in the forest for about a mile returning to base at about two in the afternoon. Now thoroughly tired and hungry, he still had to prepare his captures, a task that generally took him until five in the evening when he took tea, wrote and read, but generally was in bed by nine. [168] Scavengers and dampness are the plague of tropical butterfly collecting and Bates developed his own methods of dealing with them. In his study room at Tefé:

"...cages for drying specimens were suspended from the rafters by cords well anointed, to prevent ants from descending, with a bitter vegetable oil: rats and mice were kept from them by inverted cuyas, placed halfway down the cords."

To this could be added the difficulty of transport. Wallace's disaster returning on the brig *Helen* is widely documented elsewhere but Bates too had his disasters:

---

168 Proceedings of natural history collectors in foreign countries. Mr. H. W. Bates. *The Zoologist* 15: 5659

"As to my private collection, I find it impossible to ascertain correctly what specimens I lost in the *Mischief* and...I am very sorry to hear of the damage done to my collection at the Custom-house; that was the best box, and, in fact, I think, the very best box of butterflies I have ever sent: no one knows the days and weeks of patient search that collection cost me."[169]

It was almost certainly with these dangers in mind that Bates arranged for his final collections to be divided into three lots, each lot returning to England by separate ship.

While the transport of material between Brazil and London was clearly hazardous, it could be accomplished remarkably quickly. Writing from Santarém on 12th April 1852, Bates expresses the expectation to his London agent, Samuel Stevens that a consignment sent seven weeks previously:

"...will I hope, be at hand by this time. However, the upper Amazon was a different matter - for sailing vessels, a round trip from São Paulo de Olivença to Pará might entail a seven-month voyage."[170]

Bates financed his work by selling material through Stevens. A few entries in the British Museum of Natural History collection register indicate the sums of money involved; for instance, entry BM 1851-43 is for 55 Lepidoptera at 3/- (15 pence) each, 70 at 2/6d (12.5 pence) each, and 31 at 1/6d (7.5 pence) each. In addition, the Museum clearly had first option, much to the chagrin of other collectors like W C Hewitson, an eminent lepidopterist of his time, eagerly awaiting the opportunity to purchase some of the Bates material. Writing to Bates he said:

"I see nothing [left] of your choice things; the museum people pick them all out. They swallow up 2 or 3 of one species and one gets none."[171]

---

[169] Extracts from the correspondence of Mr. H. W. Bates. *The Zoologist* Vol 11, 1852:3322

[170] Extracts from the correspondence of Mr. H. W. Bates. *The Zoologist* Vol 11, 1852:3726

[171] Hewitson, W.C. *Illustrations of new species of exotic butterflies, selected chiefly from the collections of W. Wilson Saunders and William C Hewitson.* 5 vols. (London: John van Voost 1852-1876).

Appendix 2

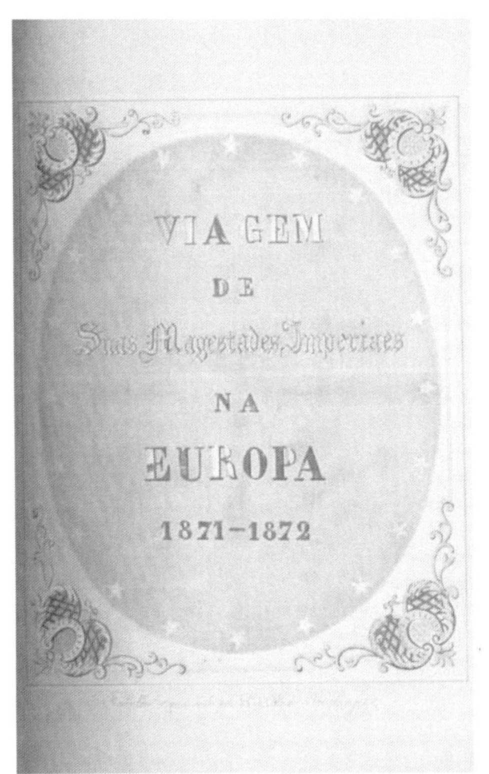

Plates 22, 23 and 24.

Hand drawn cover of a diary of the visit by The Emperor of Brazil, D Pedro II to Europe, containing details of the awards he made including Cavalier of the Imperial Order of the Rose to Bates. To the right top the obverse and to the right bottom the reverse of the medal

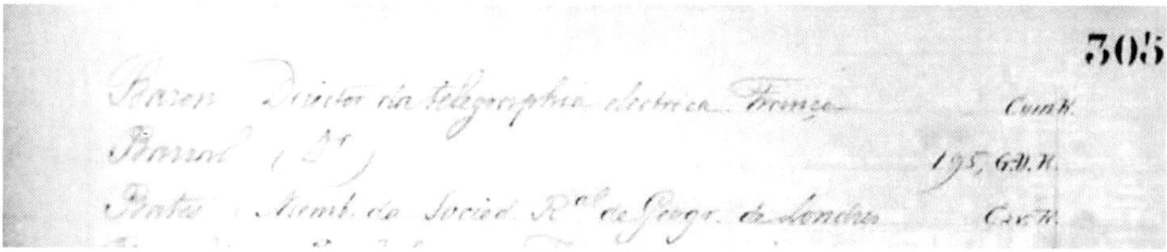

Plate 25.

Notification of Bates's medal from a private diary kept by a member of Emperor D Pedro 11 of Brazil's entourage, when he visited Europe between 1871-2.

List of recipients with Bates's name third from the top.

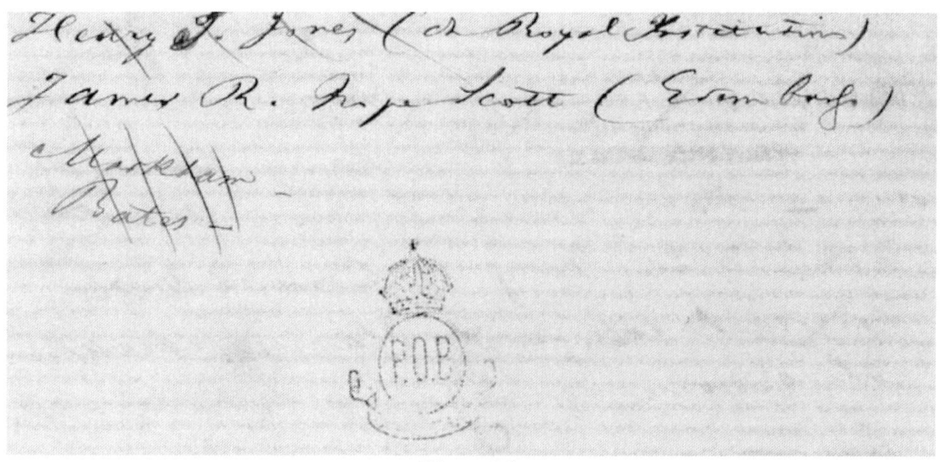
Markham
Bates

Plate 26.

Bates's signature for receipt of the medal on line four just below Markham

Appendix 3

Events in the Royal Geographical Society that were the concern of the Assistant Secretary between 1864 – 1891

1864 Grant and Speke's journey across Eastern Equatorial Africa.At this point confirmation of Bates's appointment, at a salary of £200 per annum, testified to the Society's satisfaction with his work.

1865 Captain T.G. Montgomerie's travel from the Punjab to the Karakoram Range. Samuel Baker's exploration in the interior of Africa.

1866 Dr. Isaac Hayes's expedition to the open Polar Sea.

1869 Professor A.E. Nordenskiold's Swedish expeditions to Spitzbergen.

1870 George Hayward's journey into Eastern Turkistan. Lieutenant Francis Garnier's extensive surveys of Cambodia.

1871 Much of Bates's time was taken up with matters arising from Sir Roderick Murchison's retirement. Murchison was to die the same year. For the previous forty years, Murchison, a founder member, had devoted himself to the Society; his leadership had done much to place the Royal Geographical Society in the front rank of scientific societies.

1872 Robert Shaw and his journeys in Eastern Turkistan.

1873 Henry Morton Stanley searching for Livingstone.

1874 Dr. Georg Schweinfurth's explorations in Africa.Colonel Warburton's successful journey across the previously unknown western interior of Australia.

1875 Lieutenant Weyprecht's expeditions to Spitzbergen.Lieutenant Payer's exploration in Arctic regions.

1876 Lieutenant Verney Lovett Cameron exploring Zanzibar and surveying Lake Tanganyika.John Forrest's numerous explorations in Western Australia.Captain Sir George Nares who commanded the Arctic Expedition of 1875-6 when his ships and sledge parties got nearer to the North Pole than any previous expedition.

1877 Pandit Nain Singh's journeys and surveys in Tibet and along the Upper Brahmaputra, which added much knowledge to the map of Asia.

1878 Bates was dealing with Baron von Richthofen and his extensive travels and scientific explorations in China, and Captain Henry Trotter whose work resulted in the connection of the Trigonometrical Survey of India with Russian Surveys from Siberia. The pressure of work was immense as evidenced in the fact that much of Bates's correspondence on Royal Geographical Society notepaper was actually written from his home address, even while he was on holiday.

1879 Colonel Prejevalsky's expeditions in Mongolia and the high plateau of Northern Tibet.Captain Gill's important work along the Northern frontier of Persia.

1880 Lieutenant A. Louis Palander and the Swedish Arctic Expedition in the Vega.Ernest Giles's explorations and surveys in Australia.

1881 Major Serpa Pinto's journey across Africa during which he explored 500 miles of new country. Benjamin Leigh Smith and his important discoveries along the coast of Franz-Josef Land.

1882 Dr. Gustav Nachtigal's journeys through the Eastern Sahara. Dealing with Sir John Kirk as a naturalist and second-in-command to Livingstone as well as in his position as H.M. Consul-General at Zanzibar.

1883 E. Colborne Baber and his journeys in the interior of China.
A.R. Colquhoun's journey from Canton to the Irrawadi.

1884	Dr. Julius von Haast with his extensive explorations in the Southern Island of New Zealand.
1885	Joseph Thomson and his expeditions into East Central Africa. H.E. O'Neill's exploration along the coast and into the interior of Mozambique.
1886	Major A.W. Greely who had considerably added to knowledge of the shores of the Polar Sea and the interior of Grinnell Land. At the same time Bates was dealing with Guido Cora and his important work as a cartographer.
1887	Lieutenant-Colonel T.H. Hldich's surveys of Afghanistan. Rev. G. Grenfell and his extensive explorations in the Cameroons and Congo.
1888	Clements Markham's retirement as Secretary of the Society after 25 years' service dominated events, but Bates still had to deal with Lieutenant Wiseman's explorations in Central Africa.
1889	A.D. Carey travelled nearly 5,000 miles through regions not previously visited by any Englishman.
1890	Lieutenant Younghusband and his journey from Manchuria and Peking to Kashmir, and his route surveys and topographical notes. This was the same Younghusband referred to in Bates's last letter to the Society ten days before he died.
1891	Although tiring because of age, hard work and indifferent health caused by the arduous years in the Amazons, Bates was dealing with Sir James Hector who was naturalist to the Palliser expedition and Dr. Fridtjof Nansen's crossing of the inland ice of Greenland.

Appendix 4

Bates and his family

After the birth of his daughter Alice, Bates's primary concern was his family. In a letter to Darwin in 1865 after his marriage he refers to his appointment as assistant secretary to the Royal Geographical Society and says the birth of a child with expectations of another forced upon him rational arguments for accepting the first thing offered. He became a devoted family man and deeply grieved when his daughter Alice passed away leaving an infant daughter named Dorothy Kate in 1891. Alice had married a Mr. Howard, and the infant was given into the care of Bates's other daughter Sarah, wife of William Peard, who then lived in East Finchley.

Bates had three sons, Charles Henry, (died 1901) Herbert Spencer, (died 1958) and Darwin (died 1938) (note the names) and two daughters, Alice (died 1891) and Sarah, (died 1929), who all appear to have led successful lives. After Bates died, his wife Sarah remained at Tufnell Park in the company of her son Darwin Bates. If she and his father had hoped Darwin would follow in the steps of the famous evolutionist after whom he was obviously named, they were disappointed, for his talents lay in another direction, and he made a career in the generation of electricity. After leaving University College School, in North London, Darwin became an electric engineering apprentice with Silvertown Rubber Co. Ltd. Later he joined the City of London Electric Lighting Co.Ltd., one of the supply companies involved with the first major commercial use of electric lighting for streets and public buildings in London.

In July 1894 when he was 27, and his mother had been a widow for more than two years, Darwin was married at St. John's Church, Holloway, to Alice Wilson, seven years his junior, the daughter of a local pianoforte maker, William Wilson. In the spring of 1896, Darwin left London to join the British Insulated Wire Company (eventually to become the British Insulated Cables Ltd.) at Prescot, Lancashire, as Works Manager, a job he kept until his retirement in 1933.

Darwin's claim to fame was as co-inventor of the Bates-Peard annealing furnace. His partner was G.W. Peard, who joined the company in 1899, and later married Sarah Bates, one of Bates's daughters. Until the invention of the annealing furnace, copper wire was softened by heating it in air, a method that caused oxide scale to form, which had to be removed from the wire by an acid pickling process before the wire could be used in manufacturing. The Bates-Peard annealing furnace heated the metal in a water-sealed steam-filled retort, leaving it bright and free from oxidation, a system that worked equally well with brass, nickel, silver, aluminium, and all non-ferrous metals and alloys. The result was the Bates-Peard furnace became known worldwide. Darwin now described as a person in every respect worthy of the distinction, was elected an Associate of the Institution of Electrical Engineers in January 1892 and became a Member six years later.

He was also very public minded and joined the Prince of Wales Volunteers, which was formed in 1908 and was later incorporated into the Territorial Army. Commissioned, he served with the 5th Battalion South Lancashire Regiment as second in command until World War I, when he was promoted to Colonel. In 1917, although over the age of 50 when men were not usually sent to the front line, he went to France with the men he had trained. Later he was involved with the Herefordshire Regiment before retiring from the Territorial Army. From 1921-1928 was Honorary Colonel of the 55th West Lancashire Divisional Signals. He was appointed a magistrate for West Derby in 1920 becoming a Deputy Lieutenant for the county of Lancashire in 1921. He died in December 1938.

In 1892, Sarah Bates moved from London to the house in Folkestone but remained there only briefly until in 1897 when, suffering with peritonitis, her daughter Sarah brought her back to Finchley where she died in December the same year. She was buried at her husband's side in his grave in the City of Westminster Cemetery.The other two sons, Herbert Spencer and Charles Henry became farmers in New Zealand. Charles left England when he was 18, in 1883, and was joined in 1888 by Herbert Spencer then aged 17. They thrived, became prosperous and their descendants remain there today.

Bibliography of Henry Walter Bates

Bates, H W, Notes on South American Butterflies *The Transactions of the Entomological Society of London*, Vol. V), 1859.

Bates, H W, Contributions to an Insect Fauna of the Amazon Valley, Part I: Diurnal Lepidoptera *The Transactions of the Entomological Society of London*, Vol. V), 1859 and 1861

Bates, H W, Contributions to an Insect Fauna of the Amazon Valley, Lepidoptera-Papilionidae *Journal of Entomology*, (Vol. I), 1861. 218-245.

Bates, H W, Contributions to an Insect Fauna of the Amazon Valley, Lepidoptera: Heliconiidae *The Transactions of the Linnean Society of London*, Vol. XXIII), 1862. 495-566, 2pls

Bates, H W, Contributions to an Insect Fauna of the Amazon Valley, Lepidoptera-Nymphalinae *Journal of Entomology*, (Vol. II), 1864. 175-213 & 311-346

Bates, H W, 'Contributions to an insect fauna of the Amazon valley'. *Transactions of the Entomological Society of London* (2)5: 223-228.

Bates, H W 'Proceedings of natural history collectors in foreign countries, *The Zoologist*, (1853), 11, 3726-3729.

Bates, H W, 'Proceedings of natural history collectors in foreign countries. *The Zoologist* (1853), 11, 3897-3900.

Bates, H W, 'Proceedings of Natural History Collectors in Foreign Countries', Letter to Mr. Stevens dated 10th March, 1853, from Santarém: *The Zoologist*, (1853), 11, 4113-4117

Bates, H W, 'Proceedings of natural history collectors in foreign countries. *The Zoologist* (1853), 11, 4113-4117.

Bates, H W, Proceedings of natural history collectors in foreign countries. *The Zoologist* (1857), 15, 5657-5662.

Bates, H W, 'Proceedings of natural history collectors in foreign countries. Excursion to St. Paulo, Upper Amazons'. *The Zoologist* (1858), 16, 6160-6169.

Bates, H W, *'Contributions to an insect fauna of the Amazon valley'. Transactions of the Entomological Society of London* (1861), (2)5, 335-361.

Bates, H W, 'Contributions to an insect fauna of the Amazon valley.-Lepidoptera:-Heliconiinae'. *Journal of the Proceedings of the Linnean Society of London* (1862), 6, 73-77.

Bates, H W, 'Extracts from the correspondence of Mr. H.W. Bates', *The Zoologist* (1852), 10, 3321-3324.

Bates, H W, Hints on the Collection of Objects of Natural History. *Proceedings of the Royal Geographical Society,* (1871), 16, 67-78.
Bates, H W, On the Blue-Belted Epicaliae of the Forests of the Amazons *The Entomologist's Monthly Magazine*, Vol. II, 1866.
Bates, H W, Contributions to an Insect Fauna of the Amazon Valley; Coleoptera: Longicornes *The Annals and Magazine of Natural History, Including Zoology, Botany and Geology, Vols.* VIII, IX, XII, XIII, XIV, XV, XVI, and XVII, 1861-1866.
Bates, H W, New Genera of Longicorn Coleoptera from the River Amazons *The Entomologist's Monthly Magazine, Vol. IV*, 1867.
Bates, H W, Contributions to an Insect Fauna of the Amazon Valley (Coleoptera, Prionides) *The Transactions of the Entomological Society of London for the year 1869.*
Bates, H W, Contributions to an Insect Fauna of the Amazon Valley (Coleoptera, Cerambycidae) *The Transactions of the Entomological Society of London for the Year 1870.*
Bates, H W, Notes on the Longicorn Coleoptera of Tropical America *The Annals and Magazine of Natural History, Including Zoology, Botany and Geology,* Vol. XI, 1873.
Bates, Marston and Humphrey. Eds. *The Darwin Reader.* (London: Macmillan and Co Ltd, 1957)

Further reading

Ackerman, Jennifer, Chance in the House of Fate: A Natural History of Heredity, (London: Bloomsbury Press, 2001)
Andre, Eugene, *A Naturalist in the Guianas,* (London and New York: Thomas Nelson and Sons, 1900)
Attenborough, David, *Zoo Quest to Guiana,* (London: Lutterworth Press, 1956)
Barber, Lynn, *The Heyday of Natural History*: (New York: Doubleday and Co, 1980)
Beebe, William, ed., *The Book of Naturalists: An anthology of the Best Natural History.* (Princeton, New Jersey: Princeton University Press, 1971)
Bowlby, John. *Charles Darwin: A new Life.* (New York: W. W. Norton and Co, 1991)
Brackman, Arnold C. A Delicate Arrangement, The Strange Case of Charles Darwin and Alfred Russel Wallace. (New York: Times Books, 1980)
Carroll, Sean B. *Endless Forms Most Beautiful.* (London: A Phoenix Paperback, 2007)

Caufield, Catherine, *In The Rainforest: Report from a Strange, Beautiful, Imperilled World.* (Chicago: University of Chicago Press, 1986)
Clark, Ronald W, *The Survival of Charles Darwin: A Biography of a Man and an Idea.* (London: Weidenfeld & Nicolson, 1984)
Colonel Fawcet, P H. *Exploration Fawcett.* (London: Hutchinson, 1954)
Conrad, Joseph. *Lord Jim, A Tale.* (London: Penguin books, 1993)
Cook, Laurence, & Clarke, C.A., *A Modern Aurelian*, The Linnean: Volume XIX, (2003), pages 32-42.
Darwin, Charles, *Journal of discoveries into the Geology and Natural History of the Various Countries visited by H M S. Beagle.* (London: Henry Colhurn, 1839)
Darwin, Charles, The Naturalist on the River Amazons: The Search for Evolution by Henry Walter Bates: An appreciation. *The Natural History Review*, vol III, 1863)
Darwin, Francis, ed., *The life and letters of Charles Darwin.* Two vols. (New York: Basic Books Inc, 1959).
Devries, P.J. *The Butterflies of Costa Rica and their natural history - Papilionidae, Pieridae, Nymphalidae.* (Princeton University Press: New Jersey. 1987)
Dover, Gabriel. *Dear Mr. Darwin, Letters on the Evolution of Life and Human Nature.* (London: Weidenfeld & Nicholson, 2000).
Duncan, Dayton and Burns, Ken, *Lewis and Clark, The Journey of the Corps of Discovery.* (London: Pimlico, 1997).
Edmunds, M., *Defence in Animals: A survey of anti-predator defences.* (London: Longman Press, 1974).
Engleman, Franz, *The Physiology of Insect Reproduction.* (Oxford: Permagon Press, 1970).
Everett, Susanne, *History of Slavery*, (London: Magna Books, 1978).
Ford, E. B, *Butterflies.* (London: New Naturalist: Collins, 1957).
Ford, E. B, *Ecological Genetics.* (London: Methuen and Co Ltd, 1964).
Ford, E. B, Genetic Research in the Lepidoptera, The Galton Lecture, (London: August 1939) *Annals of Eugenics*, 1940, Vol X, Part 3, pp 227-252.
Ford, E. B, *Understanding Genetics.* (London: Faber and Faber, 1979).
Freitas, A.V. L., & Oliveira, P.S., 1994. Biology and behaviour of the neotropical butterfly Eunica bechina (Nymphalidae) with special reference to larval defence against ant predation. *Journal of Research on the Lepidoptera* 31:1-11.
Freshfield, D.W. & Wharton, W.J.L., *Hints to travellers scientific and general.* Edited for the Council of the Royal Geographical Society. Sixth Edition (London: Royal Geographical Society, 1889).
Galton, F., *The Art of Travel.* (1872). (London: Phoenix Press, 1971).
Golby, J.M., Ed, *Culture & Society in Britain, 1850 – 1890.* (Oxford: OUP 1987).

Gosse, Philip, *The Squire of Walton Hall.: The Life of Charles Waterton.* (London, Toronto, Melbourne and Sydney: Cassell, 1940).
Green, Toby, *Saddled with Darwin: A journey through South America.* (London: Wiedenfeld & Nicolson, 1999).
Gullan & Cranston, *The Insects, An Outline of Entomology.* (USA: Blackwell Science, 2000).
Hartwig, Dr. G., *The Tropical World: Aspects of Man and Nature in the Tropical Regions of the Globe.* (London: Longmans, Green and Co., 1873).
Hecht, Susanna and Cockburn, Alexander, *The Fate of the Forest.* (London and New York: Verso, 1989)
Hemming, John, *Amazon Frontier, The Defeat of the Brazilian Indians.* (London: Pan Books. 2004).
Hemming, John, Die If You Must, Brazilian Indians in the Twentieth Century. (London: Pan Books, 2004).
Hemming, John, *Tree of Rivers, The Story of The Amazon.* (London: Thames & Hudson, 2008).
Howarth, Patrick, *The Year is 1851.* (London, Collins, 1951).
Imms, A. D. *A General Textbook of Entomology.* (London: Methuen & Co Ltd, 1957).
Kehlmann, Daniel, *Measuring the World.* A Novel. (London: Quercus, 2007).
Keynes, Randal, Annie's Box: Charles Darwin, his Daughter and Human Evolution. (London: Fourth Estate, 2001).
Keynes, Richard, Fossils, Finches and Fuegians: Charles Darwin's Adventures and Discoveries on the Beagle, 1832 – 1836. (London: Harper Collins, 2002).
Koster, Henry, *Travels in Brazil.* Two vols. (London: Longman, Hurst, Orme and Brown, 1817).
Leftwich, A. W, *A Dictionary of Entomology.* (London: Constable, 1976).
Linsley, E Gorton, Ed., *The Principal Contributions of Henry Walter Bates to a knowledge of the Butterflies and Longicorn Beetles of the Amazon Valley.* (New York: ARNO Press,1978).
Lloyd, Clare, *The Travelling Naturalists.* (London: Croom Helm, 1985).
McDonald, Roger, *Mr. Darwin's Shooter.* A Novel. (New Zealand: Transworld Publishers (NZ) Ltd, 1998).
McGirr, Nicola, *Nature's Connections. An Exploration of Natural History.* (London: The Natural History Museum, 2000).
Medina, José Toribo, *The Discovery of the Amazon.* (New York: Dover Publications, 1988).
Muensterberger, Werner, *Collecting, An Unruly Passion.* (Princeton: New Jersey: Princeton University Press, 1994).

Musée cantonal de Zoologie, Bibliothèque Cantanole et Universitaire, Lausanne. *Les papilions de Nabokov.* Catalogue de l'exposition. (Lausanne: Musée cantanol de Zoologie, 1993).

Nabokov, Dmitri, (New translation from the Russian,) *Nabokov's Butterflies.* by Vladimir Nabokov, (London: Allen Lane, The Penguin Press. 2000).

Neild, A.F.E. The butterflies of Venezuela. Part I; Nymphalidae 1(Limenitidinae, Apaturinae, Charaxinae). (Greenwich: Meridian Publications, 1996).

Newman, L. Hugh, *Man and Insects.* (London: Aldus Books, 1965).

O'Hara, J.E., Henry Walter Bates-his life and contributions to biology. *Archives of Natural History* 22:195-219.1995.

Owen, Denis, *Camouflage and Mimicry.* (Oxford: O. U. P., 1980).

Penz, C.M, Higher level phylogeny for the passion-vine butterflies (Nymphalidae, Heliconiinae) based on early stage and adult morphology. *Zoological Journal of the Linnean Society* 1999, 127: 277-344.

Popescu, Petru, *Amazon Beaming.* (London: Macdonald, 1991).

Raffles, Hugh, *In Amazonia: A Natural History.* (Princeton: New Jersey: Princeton University Press, 2003).

Robinson, Alex and Gardenia, Cadogan Travel guides, *The Amazon.* (London: Cadagon Guides, 2001).

Rose, Steven, *Lifelines, Biology, Freedom, and Determinism.* (London: Allen Lane, The Penguin Press, 1997).

Rothschild, Miriam, Dear Lord Rothschild: Birds, Butterflies and History. (London: Hutchinson, 1983).

Russell, Sharman, *An Obsession with Butterflies.* (London: William Heinemann, 2003).

Schappert, P.J. & Shore, J.S.. Ecology, population biology and mortality of Euptoieta hegesia Cramer (Nymphalidae) on Jamaica. *Journal of the Lepidopterists' Society* 1998 52: 9-39.

Sharp, David, A Biographical Sketch of Henry Walter Bates, FRS. *The Entomologist*, Vol. XXV, London, 1892.

Silberglied, R.E., Aiello, A. & Lamas, G.. Neotropical butterflies of the genus Anartia: Systematics, life histories and general biology (Lepidoptera: Nymphalidae). *Psyche* 1980 86: 219-260.

Simpson, John, *In the Forests of the Night: Encounters in Peru with Terrorism, Drug-Running and Military Oppression.* (London: Hutchinson, 1993)

Smith, Anthony, *Mato Grosso: Last Virgin Land.* (London: George Rainbird Ltd, 1971).

Spix, Johann Baptist von, and Martius, Karl Friedrich Philip von. *Reise in Brasilien auf Befehl Sr. Majestat Maximilian Joseph I, Königs von Baiern in den Jahren 1817 bis 1820.* (Munich, 1823-1831).
Spooner, David, *The Insect Populated Mind, How Insects have Influenced the Evolution of Consciousness.* (London: Hamilton Books, 2005).
Stecher, R.M., 1969. The Darwin-Bates letters - Correspondence between Two Nineteenth-Century Travellers and Naturalists. *Annals of Science* 25: 1-47, 96-125.
Tiller, Kate, *English Local History: An Introduction.* (Gloucestershire: Sutton Publishing. 2001).
Uglow, Jenny. *The Lunar Men.* (London: Faber and Faber. 2003).
Urquhart, F. W, *The Monarch Butterfly.* (Toronto: University of Toronto Press, 1960).
Vane-Wright, R.I, The colouration, identification and phylogeny of Nessaea butterflies (Lepidoptera: Nymphalidae). Bulletin of the British Museum (Natural History), *Entomology* 1979 38: 27-56.
Vane-Wright, R.L, Contributions to an insect fauna of the Amazon valley (Lepidoptera: Heliconiidae). *Biological Journal of the Linnean Society* 1981 16: 41-54.
Vane-Wright, R.L *Butterflies.* (London: The Natural History Museum, 2003).
Wallace, Alfred Russel, *Travels on the river Amazon and the Rio Negro*, (London: Ward Lock & Co, 1889).
Wallace, Alfred Russel. *The Malay Archipelago, The land of the Orang-Utan and the Bird of Paradise, A Narrative of Travel with Studies of Man and Nature.* (Tynron Press. 1989).
Wallace, Alfred.Russel. *My Life, A Record of Events and Opinions.* Two vols. (London: Chapman & Hall, 1906).
Waterhouse, C.O, *Insects: The history of the collection contained in the Natural History Departments of the British Museum.* vol II. (London: British Museum Natural History, 1906)
Wickler, Wolfgang, *Mimicry in plants and animals.* (London: Weidenfeld and Nicholson, 1968).
Wilson, A. N, *The Victorians.* (London: Arrow Books, 2003).
Young, A.M. Further observations on the natural history of Philaethria dido (Lepidoptera, Nymphalidae, Heliconiinae). *Journal of the New York Entomological Society* 1974 82: 30-41.
Zimmer, Carl. *Evolution: The Triumph of an Idea.* (London: William Heinemann. 2001).

Glossary

ADAPTATION: a characteristic of an organism that helps it to cope better with the conditions of its environment.
APOSEMATIC. Having a warning colouration indicating danger.
ARTIFICIAL SELECTION: human attempts to enhance natural traits by breeding selectively from those organisms that most strongly manifest those traits.
BATESIAN MIMICRY: the resemblance of a harmless species (the mimic) to another species (the model) that is endowed with poison, bad taste, or a sting or bite that makes it unpleasant to predators.
BEETLES. Coleoptera, An order of insects comprising about a quarter of a million identified species and the largest order in the animal kingdom.
BRIMSTONE. *Gonepteryx rhamni,* a large plain yellow butterfly, usually the first to be seen in the spring.
BUTTERFLIES. Lepidoptera, characterized by having clubbed antennae. During Bates's time the old system of classification divided the Lepidoptera into two groups: *Rhopalocera* or butterflies and *Heterocera* or moths. This system is still used for convenience, since it has the advantage of dividing these insects into two obvious groups easily recognized by the amateur entomologist. Butterflies are generally diurnal and are usually more brightly coloured than moths which are generally nocturnal. The largest and most brilliant butterflies are found in South America and other tropical countries. The elaborate coloured patterns of the wings are due partly to the pigments in the thousands of minute scales that cover the wings, but in addition to this there is frequently a shining blue-green iridescence due to optical interference caused by multiple transparent plates in the substance of each scale. Such colours are known as structural colours as distinct from pigments. The largest known butterfly is the New Guinea Birdwing *Triodes alexandrae,* the female of which has a wingspan of 30 cm. The smallest is the Dwarf Blue *Brephidium barberae* of South Africa with a span of about 12 mm. The families of butterflies are based largely on size, colour and habits: *Satyridae* (Browns), *Lycaenidae* (Blues and Coppers), *Pieridae* (Whites and Yellows), *Nymphalidae* (Fritillaries and Aristocrats), *Papilionidae* (Swallowtails), *Hesperiidae* (Skippers).
CARABIDAE. Ground beetles.
CHROMOSOME: structure in the nucleus of a eukaryotic cell consisting of DNA molecules that contain the genes.

CLINE: a gradual change in the frequencies of different genotypes (or phenotypes) of a species within an interbreeding population.
CRYPSIS: the resemblance, by virtue of colour or pattern, of an organism to its habitual background: camouflage.
CURARE: extracted from *Strychnos toxifera, it* is a quasi-anesthetic poison used on arrowheads by some Amazonian tribes and later used in the development of modern Western style anesthetics.
DANAIDAE. A widespread family of large migratory butterflies more common in the tropics than elsewhere.
DARWINISM: the belief that natural selection is the overwhelming causal factor in evolution.
DNA (DEOXYRIBONUCLEIC ACID): a type of nucleic acid, found in the nucleus of plant and animal cells, that contains the genetic information.
EVOLUTION: the change in groups of organisms over time, so that descendants differ from their ancestors; the change in organisms brought about by adaptation, variation and differential survival/ reproduction, by the process of natural selection.
FECUNDITY. The ability of females of a species to produce sufficient young in order to create a stable population unless some abnormal factor upsets this.
FIRE ANT. *Solonopsis,* a genus of ants from tropical South America that are a serious pest because of their irritant poisonous bite.
GENE POOL: the collective genes of a population or species.
GENE: the unit of heredity, today known to consist of a sequence of base pairs in a DNA molecule.
GENETIC DRIFT: the claim that in small populations accidents of mating can outweigh the effects of selection.
GENOTYPE: the set of genes possessed by an individual organism; often, its genetic composition at a specific locus singled out for discussion.
HABITAT: the specific type of local environment in which an organism lives.
HELICONIIDAE. A family of large brightly coloured butterflies from Central and South America. They have long slender wings, varied colouring, and a pungent smell when found in numbers.
IMAGO: the adult stage of an insect. (Final stage of metamorphosis)
ITHOMIIDAE. A family of butterflies from tropical America having a very long slender body and yellow or orange and transparent wings with thick black venation.

LARVA: the second stage in the life cycle of insects that undergo full metamorphosis; in Lepidoptera, caterpillars.
LEAF CUTTING ANTS. Tropical ants, particularly the genus *Atta* of South America also known as the Parasol Ant. These insects cut pieces of leaf, carry them above the head like a parasol and take them down into their underground nests. Here they smother them with saliva and faeces and use them as a medium on which to cultivate fungi as food for the colony. Some of their nests are as much as 10 metres across and the ants are so numerous that they will strip a large tree in a single night.
MELANISM: the deposition of melanin pigment in a structure, rendering that structure blackish. In moths, generally refers to a blackening of the wings, controlled by a single gene.
MORPH: form; a distinctive variant within a species that is characterized by genetic continuity from generation to generation, and is maintained in the species by some selective advantage.
MUTATION: a sudden change in the genetic material that determines a trait or set of traits of an organism. The mutation may be due to a change in the number of chromosomes, in the structure of a chromosome, or in an individual gene.
NATURAL SELECTION: the principal mechanism of evolutionary change according to Charles Darwin. The mechanism whereby advantageous heritable traits, which enhance the ability of an organism to survive or reproduce, are more likely than disadvantageous traits to be passed on to future generations.
PHENOTYPE: the physical and behavioural properties of an organism, manifested throughout its life.
PHEROMONE: a chemical substance secreted or released by an organism that influences the behaviour of other individuals of the species. For example, some pheromones function as sexual attractants.
POLYMORPHISM (GENETIC): the occurrence in a single interbreeding population of two or more sharply distinct heritable forms, the rarest of which is too frequent to be maintained simply by recurrent mutations.
POPULATION: a group of organisms of the same species living and breeding together in the same place.
PUPA: the life-cycle stage between larva and adult (imago) in insects having complete metamorphosis.
RECESSIVE: a recessive allele is expressed in the phenotype of homozygotes and not in that of heterozygotes. Opposite of dominant.

SELECTIONISTS: evolutionists who hold that most evolutionary change results from the action of natural selection on small heritable differences.
SPECIATION: the process whereby new species form.
SPECIES: groups of interbreeding or potentially interbreeding populations that are reproductively isolated from other such groups.
SUGARING: a technique for attracting certain moths, usually by applying a sugary and slightly alcoholic mixture to tree trunks.
SYSTEMATICS: the scientific theory behind the classification of organisms.
TAXONOMY: the study of the classification of organisms.

Index

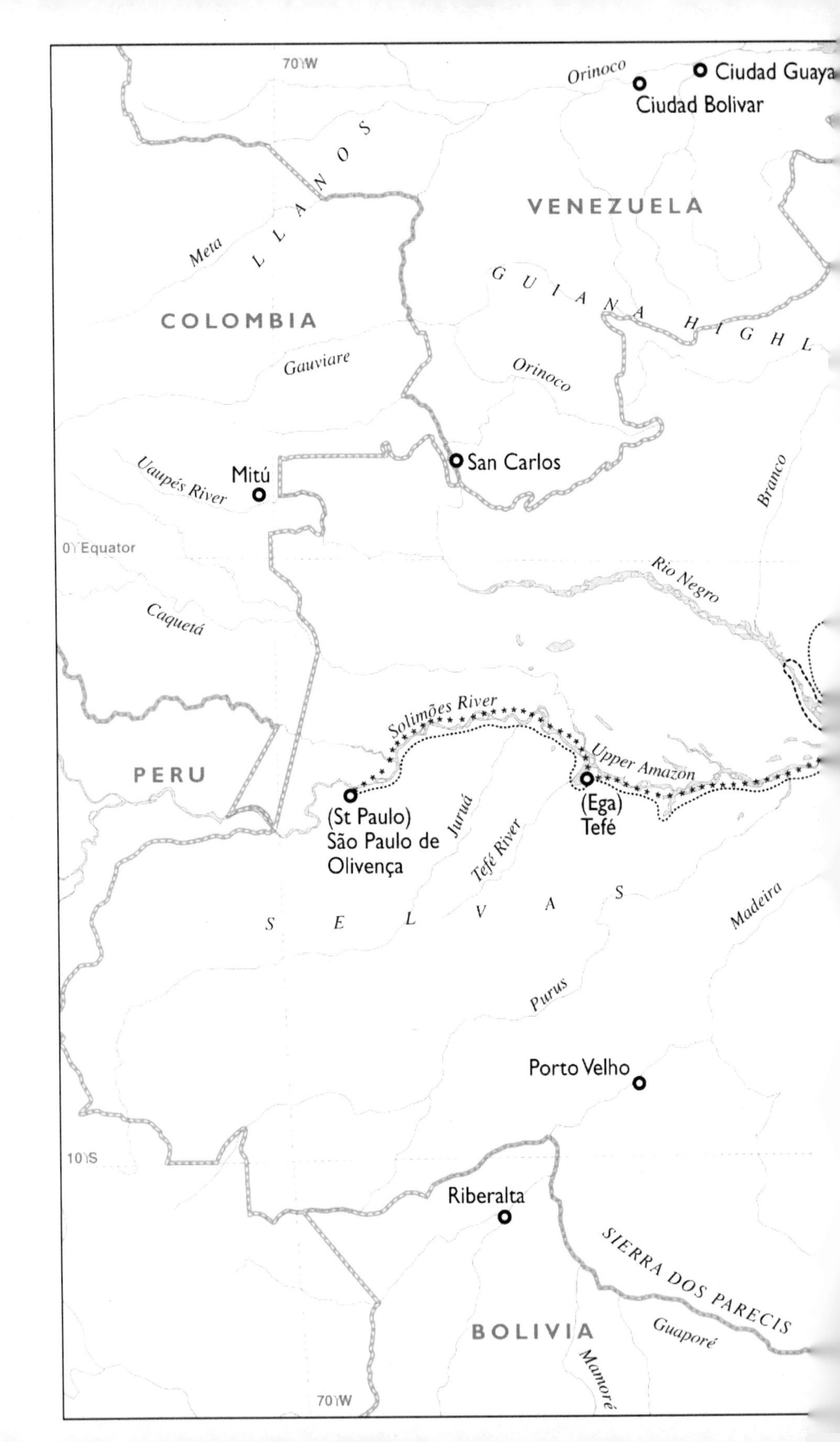

70°W
Orinoco
Ciudad Guaya
Ciudad Bolivar
LLANOS
VENEZUELA
Meta
GUIANA HIGHL
COLOMBIA
Gauviare
Orinoco
San Carlos
Mitú
Uaupés River
Branco
0° Equator
Rio Negro
Caquetá
Solimões River
Upper Amazon
PERU
(St Paulo)
São Paulo de
Olivença
Juruá
(Ega)
Tefé
Tefé River
SELVAS
Madeira
Purus
Porto Velho
10°S
Riberalta
SIERRA DOS PARECIS
Guaporé
BOLIVIA
Mamoré
70°W